About Island Press

Island Press is the only nonprofit organization in the United States whose principal purpose is the publication of books on environmental issues and natural resource management. We provide solutions-oriented information to professionals, public officials, business and community leaders, and concerned citizens who are shaping responses to environmental problems.

In 1994, Island Press celebrated its tenth anniversary as the leading provider of timely and practical books that take a multidisciplinary approach to critical environmental concerns. Our growing list of titles reflects our commitment to bringing the best of an expanding body of literature to the environmental community throughout North America and the world.

Support for Island Press is provided by Apple Computer, Inc., The Bullitt Foundation, The Geraldine R. Dodge Foundation, The Energy Foundation, The Ford Foundation, The W. Alton Jones Foundation, The Lyndhurst Foundation, The John D. and Catherine T. MacArthur Foundation, The Andrew W. Mellon Foundation, The Joyce Mertz-Gilmore Foundation, The National Fish and Wildlife Foundation, The Pew Charitable Trusts, The Pew Global Stewardship Initiative, The Rockefeller Philanthropic Collaborative, Inc., and individual donors.

Certification of Forest Products

Certification of Forest Products

Issues and Perspectives

Edited by

| Virgílio M. Viana | Jamison Ervin | Richard Z. Donovan | Chris Elliott | Henry Gholz |

ISLAND PRESS

Washington, D.C. ■ Covelo, California

Grateful acknowledgment is made for permission to reprint the following previously copyrighted material.

A version of Chapter 9 originally appeared in O'Hara, J., Endara, M., Wong, T., Hopkins, C., and Maykish, P. (eds.). 1994. *Timber Certification: Implications for Tropical Forest Management.* Proceedings from a conference hosted by the student chapter of the International Society of Tropical Foresters, Yale School of Forestry and Environmental Studies, New Haven, Connecticut.

Chapter 10, in a different, longer version, originally appeared in Baharuddin, H., and Simula, M. 1994. *Certification Schemes for All Timber and Timber Products.* Yokohama, Japan: ITTO. "Labor's View," found on pp. 218–219, and "Key Issues on Certification of All Timber and Timber Products," found on pp. 239–241, are from appendixes of the same publication. Reprinted by permission of ITTO.

"The Society of American Foresters," found on pp. 193–194, is reprinted form *SAF Forest Management Certification Report,* 1995, published by the Society of American Foresters, 5400 Grosvenor Lane, Bethesda, Maryland. Not for further reproduction.

"Menominee Tribal Enterprises," found on pp. 197–198, originally appeared in "Case Studies of Community-based Forestry Enterprises in the Americas," presented at the symposium Forestry in the Americas, Community-based Management and Sustainability, University of Wisconsin–Madison, Feb. 3–4, 1995. Madison: Institute for Environmental Studies, Land Tenure Center. Copyright © 1995.

Library of Congress Cataloging-in-Publication Data

Certification of forest products: issues and perspectives / edited by
    Virgílio Viana . . . [et al.].
        p.   cm.
    Includes bibliographical references and index.
    ISBN 1-55963-493-6 (cloth).—ISBN 1-55963-494-4 (paper)
    1. Forest management—Standards. 2. Forest products—
Certification. I. Viana, Virgílio.
SD387.S69C47   1996
634.9'2'0218—dc20   96-11219
                                                                 CIP

Printed on recycled, acid-free paper

Manufactured in the United States of America

10 9 8 7 6 5 4 3 2 1

Contents

Preface

Certification of forest products is a new and exciting phenomenon. Its future is still uncertain, but it could have a significant impact on forest practices around the world in coming years. Although the concept of certification in other fields, such as organic agriculture, has existed for decades, it was only in 1990 that the first forest operation was certified. In the last few years, policy makers, forestry professionals, environmentalists, and social activists have begun paying increased attention to forest product certification. A number of recent workshops, symposia, and other meetings have sparked even greater interest in forest product certification worldwide.

Forest product certification is a reality. Nearly one and a half million cubic meters of certified timber is produced annually, and at the time this book went to press eight certifiers were operational. While a great deal of information has been generated through these activities, access to information about forest product certification is limited and uneven across geographical regions and stakeholder groups. There is a marked scarcity of published information on the history, current status, and potential impact of forest product certification.

The idea for this book germinated during a working group meeting of the Forest Stewardship Council (FSC) in the summer of 1994. The editors had been closely involved with forest product certification and were familiar with many of the current issues. While four of the five editors have contributed to the formation of the Forest Stewardship Council, this is not a book specifically about that organization. Nor is it intended to be a handbook on certification. The intention of the editors is to provide objective information to help the reader understand the certification process and to critically analyze its potential and limitations. This book is aimed at a wide audience, including policy makers, business executives, forest owners and managers, certifiers, consultants, researchers, nongovernmental organizations, and students.

The editors held two writing workshops in 1995-one in Piracicaba, Brazil, the other in Bolton, Vermont. Each editor was responsible for writing, collaborating, and/or editing one or more chapters. Chapter authors other than the editors were invited to contribute based on their specific knowledge and experience. Chapter authors were provided with general terms of reference, including the length, objectives, key points to be raised, and an outline of the entire book. Authors of the perspectives in Part III were chosen to reflect a range of opinions. Contributors to Part III were asked simply to write a short piece describing their views on certification.

Acknowledgments

Many people contributed to this book. We would like to thank our host institution in Brazil, the Department of Forest Sciences of University of São Paulo, and the owners of the Black Bear Inn, Sally and Denis Turpin, who warmly welcomed our stay in Vermont. In Brazil we also had the invaluable organizational support of Tasso Rezende, of IMAFLORA. The World Wide Fund for Nature International provided financial assistance to this project. Henry Gholz was a Science and Engineering Diplomacy Fellow of the American Association for the Advancement of Science at the United States Agency for International Development Center for the Environment during his participation in this project. Additional thanks to the Fundação de Apoio à Pesquisa do Estado de São Paulo and the Tropical Conservation and Development Program of the University of Florida for their support for a sabbatical leave for Virgílio M. Viana at the School of Forest Resources and Conservation of the University of Florida. The following individuals helped in different ways: Per Rosenberg, Markku Simula, Odette Jonkers, Francis Sullivan, Andrew Poynter, Alan Pierce, Arlin Hackman, Neil Byron, Carole Saint-Laurent, Susan Morgan, Jack Putz, Bruce Cabarle, Timothy Synnott, Charles Arden-Clarke, Jean-Paul Jeanrenaud, Matthew Wenban-Smith, Peter Polshek, C. Chapman, G. Williams-Linera, and Clyde Kiker. To all the above individuals and institutions: Thank you! All errors and omissions remain the responsibility of the editors and authors.

The Editors

Chapter 1

Introduction

Chris Elliott and Richard Z. Donovan

Over the past 10 years, forest conservation has become an increasingly high-priority issue for both policy makers and the general public throughout the world. Initial concern focused on tropical forests and the activities of the timber industry. In the mid-1980s, two international initiatives were launched to lessen the industry's impact on tropical forests: the Tropical Forestry Action Program (TFAP) and the International Tropical Timber Organization (ITTO).

In 1992, forest conservation was featured prominently on the agenda at the United Nations Conference on Environment and Development (UNCED) held in Rio de Janeiro. The discussions were highly controversial and the conference produced nothing more substantial than a set of nonbinding "Forest Principles." However, an important achievement was the inclusion of temperate and boreal forests, along with tropical forests, on the international agenda. Since UNCED, the focus of the international policy debate has widened to include temperate and boreal forests, and a wide range of international forest initiatives (such as the 1993 Ministerial Conference on the Protection of Forests in Europe) have been launched to deal with temperate and boreal forest issues.

Certification is the process in which a forest owner voluntarily requests an independent certification body to inspect his or her forest land. The certifier visits the forest site and determines whether the management meets clearly defined standards and criteria. The certification process may also include an audit of the forest product from the log yard to the final point of sale, provided the forest owner wishes to sell the product as certified. This process, known as forest product certification, allows consumers to identify products that come from well-managed forests.

Initially, certification was aimed at tropical timber. Gradually its breadth increased to incorporate all forest types, including boreal and temperate forests. Certification was originally promoted by conservation nongovernmental organizations (NGOs), but in the last three years there has been an increasing level of interest in certification on the part of forest industries in several major timber-producing countries, including Brazil, Indonesia, Sweden, and Canada.

Deforestation and Forest Degradation

In most temperate zones, the area covered by forest is currently stable or even expanding. However, as more and more old-growth forests are being replaced by plantations, the quality of this forest cover has declined. This poses a serious threat to biodiversity. In Sweden, for example, forestry activities are thought to threaten over 300 living species. Forest decline affects 27 percent of Europe's broadleaved trees and 14 percent of its conifers (Dudley 1992).

In the tropics, deforestation is on the increase. Statistics from the United Nations Food and Agriculture Organization (FAO) reveal that deforestation rates in the tropics have increased from 11.3 million hectares in 1980 to 15.4 million hectares in 1990. At current rates, the world is losing 0.8 percent of its tropical forests each year. Most severely affected is Southeast Asia, where forests are disappearing at an annual rate of 1.6 percent in continental areas (FAO 1993).

Deforestation also has serious social, ecological, and economic implications. It is not easy to quantify these, but they can be substantial and include loss of biodiversity and homelands for forest dwellers. Economic losses are sometimes easier to document: For example Thailand, traditionally a timber-exporting nation, had by 1992 become the world's largest importer of tropical sawnwood because of the depletion of its own forests.

Although the fate of most deforested areas tends to be the same—almost all end up being turned into farmland—deforestation occurs for many reasons and in many different ways. The issue is both complex and delicate.

The role of the timber industry and the international timber trade is particularly controversial. A report for the International Tropical Timber Organization (ITTO) argued that "The [international timber] trade is not a major source of tropical deforestation," claiming that conversion of forests to other uses such as agriculture is a more significant factor, and pointing out that an increasing proportion of the tropical timber harvested in producer countries is consumed domestically. The report states: "Only 6 percent of tropical non-coniferous roundwood production enters the international trade" (LEEC 1993). WWF, on the other hand, maintains that "logging of native and old-growth forests is currently the single most damaging factor affecting temperate forests" (Dudley 1992). The traditional focus on sustaining timber production rather than on the multiple functions of forests has led to a reductionist approach to forest management in many countries, with a consequent loss of forest quality.

Despite these wide differences of opinion about the causes of forest degradation, considerable progress has been made in national and international discussions on the links between environmental protection and sustainable development. Most countries are now trying to work out how to implement sustainable forest management. A recent ITTO publication notes: "There is no more debate on the 'why and wherefore' of conserving ecosystems and forests and promoting economic development albeit sustainable development. Discussion and dialogue has moved on to a matter of modalities, practicalities, criteria, technical details, etc." (ITTO 1994)

An Unsolved Problem

The failure of international initiatives to halt deforestation and promote improved forest management—in many tropical countries the situation is worsening—has prompted people to question the appropriateness of the approaches being used.

One problem is that approaches have not been consistent. While development aid agencies were increasing their spending on forestry in the 1980s (TFAP doubled international development assistance funding in the forest sector between 1985 and 1990), some of the same institutions in developing countries that were receiving increased funds were simultaneously being subjected to drastic budget cuts as part of International Monetary Fund structural adjustment programs.

Also, some conservation NGOs were promoting bans and boycotts of tropical timber as a mechanism to reduce demand for tropical timber. As a result, municipalities in the Netherlands, Germany, and the United States banned the use of tropical timber in government-funded construction. Boycotts of tropical timber, and a boycott of some Canadian timber companies promoted for some time by Greenpeace, have had a major psychological impact on the timber trade. As a result, some forest products companies may be more receptive to certification than they might otherwise have been. The impact of boycotts on timber consumption is difficult to assess, but anecdotal evidence suggests that there have been significant effects in Northern Europe.

Recently there has been a reaction against boycotts, with questions being raised about their effectiveness and fairness. In the wake of these attempts to solve the world's forest problems, governments, NGOs, and the private sector are now reviewing their policies and approaches to forestry and making a concerted effort to come up with new and more effective tools, including certification.

Certification was developed as an alternative to the perceived inefficiency of international initiatives, government policies, and boycotts in reducing deforestation and promoting sustainable forest management. It is being used as a "soft policy tool" by NGOs and the private sector to reach environmental goals through the provision of market incentives. The ultimate objective of certification is to provide an economic incentive to forest managers voluntarily interested in promoting forest management practices that are in accordance with principles of sustainable development.

The Policy Makers

In the past, policies concerning forestry or the environment were generally made and implemented by governments, while NGOs and the private sector sought to influence those governments. Recently, however, that role has changed (Chapter 8). NGOs and the private sector have started to look for policy instruments that they can develop and implement themselves. This sometimes involves collaboration between NGOs and the private sector, with the government taking a monitoring and supporting role.

In many countries the general public is also now taking a more active role in shaping policy. Individuals who are interested in particular environmental or social issues are tending to modify their consumption patterns to promote certain objectives—avoiding buying certain products, for example, and making wider use of bicycles and public transport.

There is also an increased recognition of the potential of market-based policy instruments rather than exclusive reliance on "command and control" policy instruments. For better or for worse, the market is seen to have an increasing influence on modern societies. One rather extreme opinion supporting this viewpoint has been expressed by O'Laoire (1995):

Traditional institutions and structures that previously acted as sources of security and comfort for society have notably declined in western society. This is perhaps most notable where the role of family and community has diminished. . . . With the falling away of family structures, community structures, and other institutions the individual, by default, has become more reliant on the market as the ultimate social institution to which he or she looks, not only for physical needs but also for social, cultural and even spiritual needs—a whole spectrum of lifestyle needs of which the marketing profession is well aware. This dependence on the market paradoxically increases the anxiety level. It is therefore not surprising that increasingly in the welfare societies of the first world, price is no longer the primary determinant of choice. A complex web of assurances is required from the market, ranging from ethical, animal welfare, health, hygiene and environment.

Current Status of Certification

It can be seen from the preceding pages that at the international policy level the conditions are ripe for certification to emerge as a new policy instrument. Already the first signs of a viable market-driven certification system are emerging. The first certified forest operation was an Indonesian teak plantation, certified by the Smart Wood program in 1990. In 1992, the Woodworkers Alliance for Rainforest Protection proposed the establishment of the Forest Stewardship Council (FSC). After a meeting in Washington, D.C. later that year, an interim FSC board was elected to carry out extensive consultations and organized the FSC founding assembly in October 1993. The first certifiers were accredited by FSC in 1995. In the same year, the Canadian Standards Association (CSA) made a proposal that the International Organization for Standardization (ISO) develop standards for sustainable forest management.

Eight certifiers were operational in the United States, the United Kingdom, and the Netherlands as of January 1996. Many more are preparing to start work in countries such as Brazil, Mexico, Costa Rica, Guatemala, Indonesia, and Ecuador. The policy debate on certification and national certification standards has been active in countries such as Sweden, Indonesia, Bolivia, Switzerland, Canada, and Australia. Throughout this book examples of certifiers' programs and national consultations are used for illustrative purposes.

Certification is a rapidly evolving field. While any attempt to provide a detailed picture of the current status of certification may soon be outdated, the authors have included an overview of the certified forests (Table 1.1) and the operational certifiers (Table 1.2) for comparative purposes.

Table 1.1
Certified forests[a]

Country	Certified Forest
Brazil	Amacol Ltd., Portel, Pará (c/o) Larson Wood Products, Eugene, Oregon 59,000 ha 25,000 cu m Rainforest Alliance.
Costa Rica	Bosque Carrillo Puerto S.A.: plantations. Eco-Rating International, of Zurich. 1992.
	Tropical American Tree Farms, Campo Real and Santo Domingo, Baru, Perez Zeldón. Young plantations, not yet in production. Rainforest Alliance. 1336 ha total area. 595 ha due for productive plantations.
	Portico S.A. Supplies Royal Mahogany Products. SCS.
Honduras	Proyecto Desarrollo Bosque Latifoliado, La Ceiba. 12 certified communities, total perhaps 30,000 ha Rainforest Alliance. 25,000 ha 100,000 bd.ft/yr.
Indonesia	Peru Perhutani Forest Plantation System, of the State Forestry Corporation, Java. Rainforest Alliance. Plantations of teak, mahogany, rosewood, pine. 2,331,500 total ha, 730,000 cubic meters available.
Malaysia	Guthrie Estates: Yong Peng and Sungai Labis Estates of Kumpulan Guthrie BHD, supplied to Chindwell Doors, Johor and B&Q UK. SGS-Forestry 3284 + 1603 ha of estate land. 1132 + 324 ha of rubber plantations.
Mexico	Plan Estatal Forestal: seven ejidos. Natural Forest. SCS and Rainforest Alliance.
Papua New Guinea	Bainings Community-Based Ecoforestry Project. 125,000 ha of production forest. SGS-Forestry. FSC inspection Feb. 1995.
	Masurina. Natural communal forests. Rainforest Alliance.
United Kingdom	Dartington Home Woods, Devon. 90 ha Soil Association. Plantations
	Pengelli Forest, Dyfed, Wales. 65 ha Soil Association. Old natural forest.
United States of America	Allan Branscomb, Eugene, Oregon, and one other forest in Oregon certified by Rogue Institute: 48 and 160 ha.
	California: two small forests certified by Institute for Sustainable Forestry.
	Collins Pine Company: the Collins Amanor Forest, Chester County, CA 39,000 ha pine forest. SCS.

Kane Hardwood Division of Collins Pine Co., Kane Pennsylvania. SCS. 46,500 ha.

Keweenaw Land Association, Ironwood, Michigan. Rainforest Alliance. 62,000 ha mixed hardwood forest.

Menominee Tribal Enterprises, Keshena, WI. Indigenous semi-natural forest. SCS and Rainforest Alliance.

Oregon State: two forests certified by Rogue Institute: 48 and 160 ha.

Seven Islands Land Co.: Pingree Family Ownership. Bangor, Maine. 390,000-ha forest. SCS.

[a]The forests listed have been independently certified by existing certification organizations, though not necessarily according to the FSC Principles and Criteria.

This book covers a range of issues. Part I introduces the concept of certification. This process, which involves an independent assessment of a specific forest management operation, evaluates ecological, social economic, and silvicultural practices according to agreed-upon forest management standards. Because these standards provide the basis for on-the-ground certification (Chapter 5), they are often the source of debate and controversy. Some standards, such as those promoted by the Forest Stewardship Council, are performance-based and focus primarily on the forest management practices themselves. Other standards, such as those promoted by the International Organization for Standardization, are systems-based and focus primarily on the management systems involved (Chapters 3 and 4). Part I outlines a process for developing consensus on forest management standards through a consultative process (Chapter 2).

Certification also involves monitoring and inspecting every link between the forest and the retail product. This chain-of-custody auditing may include inspections at the forest operation, the sort yard, the sawmill, the primary and secondary manufacturer, the importer and exporter, the wholesaler, and the retailer. This process brings to certification its own challenges and opportunities (Chapter 6).

Part II reviews a number of key issues raised by certification. As mentioned earlier, some environmental organizations have begun to take an active role in promoting certification (Chapter 8). The potential and limitations of certification as a policy tool (Chapter 7) and as a catalyst for change (Chapter 9) are also discussed.

While the number of certified forests is growing rapidly, demand for certified products currently exceeds supply. The "1995 plus group" in the United

Table 1.2
Certification organizations[a]

Country	Organization
Brazil	SBS-Cerflor: Rubens Garlipp, São Paulo, Fax 55 11 819 1771.
	IMAFLORA: Tasso Rezende de Azevedo, Tel/Fax 55 194 33 0234 or 22 6253.
Canada	Alberta Forest products Association Forest Care Certification Program
	Ecoforestry Institute Society, PO Box 5783, station B, Victoria, BC, V8R 6S8. Tel and Fax 604 598 8529. Robert W. Dixon. Doug Patterson.
	Silva Forest Foundation, PO Box 9, Slocan Park, British Columbia, V0G 2E0. S. Whitaker. Susan Hammond.
Chile	Fundación Chile
	INFOR: Instituto Forestal, Santiago.
Costa Rica	Fundación Ambio RNT: Recursos Naturales Tropicales.
Finland	Indufor Oy, Kulmakatu 5 B 7, FIN-00170, Helsinki.. Tel 358 0 135 2233. Fax 2552. Mr. Markku Simula.
Germany	Black Forest Certification Initiative
Indonesia	Indonesian Ecolabeling Institute
Kenya	Environmental Liaison Center. Mr. Ranil Senanayake. NTFPs.
Malaysia	National Timber Certification Committee
Netherlands	Institute for Certification of Timber, Mennonietenweg 10, 6702 AD Wageningen. Mr. Karel O. Pavlicek.
	SKAL, Stationsplein 5, PO Box 384, 8000 AF Zwolle. Tel 31 38 226 866. Fax 213 063. Member of IROAM. Mr. Jan-Willem Heezen.
Mexico	CCMSS: Consejo Civil Mexicano para Silvicultura Sostensible, México D.F.
Solomon Islands	Soltrust/Iumetugetha Holgings, PO Box 748, Honiara. Tel 677 30948. Fax 677 30468. Anthony Carmel
Switzerland	Eco-Rating International, Ackersteinstrasse 45, 8049 Zurich. Tel and Fax 41 1 342 1553

United Kingdom	TRADA Certification, Stocking Lane, Hughenden Valley, High Wycombe, Bucks HP14 4NU. Dr. C. Gill.
United States of America	Entela, Minnesota.
	Forest Trust, New Mexico.
	Institute for Sustainable Forestry: Pacific Certified Ecological Forest Products, PO Box 1580, Redway CA 95560. Tel 1 707 923 4719. Fax 923 3241. Tracy Katelman. Walter Smith.
	Northeast Natural Resources Center, National Wildlife Federation, 18 Baldwin Street, Montpelier, VT 05602. Tel 1 802 229 0650. Fax 4532. Eric Palola.
	Rogue Institute for Ecology and Economy, 1745 W. 15th Street, Eugene, Oregon 9744402. Tel 1 503 346 3653. Fax 346 2040. PO Box 3213, Ashland OR 97520 Tel 503 402 6031. Fax 7282. Brett Ken Cairn.
	Sigurd Olsen Environmental Institute, Northland College, Ashland, WI 54806. Tel 1 715 682 1223 & 373 2988. Fax 373 2938. Bob Brander. Kathleen Lidfors.
	Zobel Forestry Associates, PO Box 37398, Raleigh, North Carolina 27627. Tel 1 919 469 5054. Fax 467 0329.

[a]This is a provisional list of operational certifiers and others who have expressed interest in certification or made enquiries.

Kingdom, which is a voluntarily formed group of retail stores with sales exceeding $2 billion per year, have publicly committed to rapidly introducing certified products. Part II looks at some of the market and economic considerations (Chapter 10), as well as some of the unintended consequences of certification, such as wood substitution (Chapter 11). This part also explores key issues and challenges in certifying nontimber forest products, such as resins, medicinals, fruits, and vines (Chapter 12).

In many national and international forestry debates there has been a shift from a traditional focus on sustained-yield timber production to an integrated approach that seeks to reconcile wood production with watershed protection, biodiversity and soil conservation, equitable social benefits, and respect for local rights. However, this approach calls for new ecological, social, and economic research tools. The final chapter of Part II examines some of the research needs and information gaps for the broad application of certification (Chapter 13).

Part III includes a variety of perspectives on certification—opinions from conservation NGOs, forestry professionals, community groups, businesses, certifiers, and regional, national, and government perspectives.

Forest product certification is an evolving concept. Its impact in the coming years remains uncertain. Some have adopted certification enthusiastically. Others actively oppose it. Still others remain undecided. Most, however, would agree that in one form or another, forest product certification is here to stay.

References

Dudley, Nigel. 1992. "Forests in Trouble; a Review of the Status of Temperate Forests Worldwide." WWF International, Gland, Switzerland. 260 pp.

Food and Agriculture Organization of the United Nations, FAO. 1993. "Forest Resources Assessment." Rome, Italy.

International Tropical Timber Organization, ITTO. 1994. "Report of the Working Party on Certification of All Timber and Timber Products." ITTO, Yokohama, Japan. 200 pp.

London Environmental Economics Centre (LEEC). 1993. "The Economic Linkages Between the International Trade in Tropical Timber and the Sustainable Management of Tropical Forests." Report to ITTO. LEEC, London, U.K. 132 pp.

O'Laoire, D. 1995. "Concerns of the Small and Medium Enterprise Sector." Paper presented to International Organization for Standardization/CASCO Workshop. Geneva, Switzerland.

Part I

The Certification Process

In theory, forest product certification simply entails an independent assessment of the management practices of a particular forest management unit, and an audit of the product from stump to consumer. In practice, though, certification is far more complex. This part outlines the various elements of a certification assessment.

The basis of all certification is forest management standards. If these standards are to meet the needs of a diversity of stakeholders, they must be developed through an inclusive, consensus-building process. Chapter 2 describes some of the challenges and pitfalls of developing standards through a consultative process. Various perspectives, key issues, and limitations of the consultative process are presented. Chapter 3 describes the forest management standards themselves, including various types and definitions of standards, their purpose, and other key aspects.

An important aspect of forest product certification is credibility. Unless a set of global principles is agreed upon, certification systems may vary wildly from one forest to the next. Accreditation is an independent system of evaluating and monitoring certifiers, according to a set of internationally accepted principles. This process ensures that certification is consistent, credible, and reliable. Chapter 4 describes some of the elements of accreditation and cites an example of one existing accreditor, the Forest Stewardship Council.

Chapter 5 describes certification practices and methods of assessing forest operations. Issues include certification procedures for evaluating and categorizing natural forests and tree plantations. Chapter 6 describes the process of monitoring a forest product from the stump to the consumer, a process called chain of custody. Various technologies and methods are described at each stage. Chapter 6 also includes the constraints, potential solutions, and benefits of monitoring the chain of custody.

Chapter 2

The Consultative Process

Jamison Ervin

There is a growing recognition that sustainability can be achieved only with the participation and consent of those for whom the benefits of sustainable development are intended. Since the 1992 United Nations Conference on Environment and Development in Rio de Janeiro, it has become widely accepted that consultation and consensus building should be an integral part in the design and implementation of sustainable development. Statements to the effect that adequate consultation "provides a forum and process through which participants can build a consensus on the sustainable development of their region" (IUCN, UNEP, and WWF 1991) have become commonplace, although implementation of these approaches has been uneven in practice.

Consultation may not always be a practical or desirable decision-making tool. However, in the context of forest product certification, a broad consultative process may serve to broaden public support for, and understanding of, the certification process. This process can also help ensure that forest management standards reflect the needs and interests of affected stakeholders and can address a broad range of interests and concerns.

This chapter proposes a framework for analyzing consultative processes, presenting three examples where such a process was used in the development of forest management standards. Also presented are key elements and poten-

tial limitations of a consultative process, and questions are raised for designers and implementers of similar processes.

Four Organizational Frameworks

Defining the consultative process can be deceptively difficult. Just as organizations can be "multiple realities, subject to multiple interpretations," so too
are organizational decision-making procedures "complex, surprising, decep-
tive, and ambiguous" (Bolman and Deal 1990). Perhaps no other decision-
making process is as rich and complex as the consultative process.

To better understand the consultative process, it is tempting to reduce the
term to its elements, as simply "the act of consulting in order to obtain information or advice or to influence a particular decision" (Dodge 1978), or as
"the means by which anybody concerned—communities, governments, industry, other interest groups and individuals—can participate in developing a
strategy" (IUCN, UNEP, and WWF 1991).

A more insightful approach, however, is to first understand some of the objectives and interpretations underlying the consultative process. Bolman and
Deal (1990) provide four frameworks for analyzing organizations. These
frameworks, (1) structural, (2) human resource, (3) political, and (4) symbolic, are also useful tools for analyzing the consultative process. (See Table
2.1.) By examining the consultative process from various perspectives one can
better understand some of the multiple interpretations and expectations attached to this decision-making process.

A consultative process can have elements from all of these frameworks. It
can be a simple mechanism for gathering information to rationally design an
organization, draft a document, or formulate a policy. It can be a technique
for drawing people in, to create consensus and build networks and strengthen
relationships, and to best utilize individual energy and talent. It can be a dynamic forum for positioning, bargaining, negotiating, and for gaining or losing power. And it can even be seen as a symbol of participatory democracy—
of a sustainable future based on decision making that is transparent, inclusive,
and equitable.

There are a number of issues to consider when designing and implementing a consultative process. Some of these issues are based in part on the expectations, interpretations, and understandings attached to the consultative
process. Each of the four frameworks provides a different focus and a different perspective.

From a structural perspective, the design of the process, including the use
of an iterative learning process, the scale of the process, the particular techniques employed, and mechanisms for future input are important considera-

Table 2.1
Framework for analysis

Aspects	Structural Approach	Human Resources Approach	Political Approach	Symbolic Approach
Focus	On the institutional structures of an organization	On the human needs and personal resources within an organization	Allocation of resources and power within an organization	Organization's corporate culture: symbols, myths, imagery
Definition of Planning	A strategy for clarifying objectives	An opportunity to promote participation	A forum for airing conflicts and stating positions	Clarifying ambiguity, reducing chaos
Decision Making	The rational outcome of careful planning	A process for inducing commitment to the outcome	An opportunity for exercising or gaining power	A shared exercise in making meaning
Expectations of Consultative Process	A mechanism for gathering missing information	A mechanism to improve working relationships	A forum in which individuals and coalitions can negotiate	A symbol for something larger (north/south equity, sustainable development, etc.)

tions. From a human resource perspective, the diversity of stakeholders and the extent of participation are particularly relevant. From a political perspective, authentic representation of stakeholder groups and equitable decision-making procedures are key issues. From a symbolic perspective, the decision-making style as well as the transparency, accountability, and ownership of the process are fundamental considerations.

At any one time, participants of a consultative process will interpret the process through any of these four frameworks. At meetings, for example, there may be misunderstandings and confusion as participants may act according to their particular framework interpretation. Some may view a meeting as a chance to bring thoughtful people together to create new information, while others may view any meeting as a chance to negotiate their positions and influence others. A successful consultative process, therefore, will include elements from each approach.

Examples of Consultative Processes in the Development of Standards

The following three examples (Lake States, Swedish, and Forest Stewardship Council) are drawn from consultative processes conducted from the period 1992 through 1995. To some extent, all three processes are ongoing. These cases reflect some of the distinct differences inherent in the design and execution of a consultative process, including differing approaches, processes, and scales. The Lake States process was intended for a relatively small scale, including three states in the Great Lakes Region of the United States. The Swedish process was designed for all of Sweden. The Forest Stewardship Council process was intended to cover all forest types in all countries.

Lake States Consultative Process

The Sigurd Olson Environmental Institute, an affiliate of Northland College, recognized the importance of certification and the need for regional certification standards. They decided to contribute to the development of standards for the Lake States Region (Wisconsin, Minnesota, and Michigan). They later decided to become a formal certification partner of the Smart Wood program.

Process

1994   Sigurd Olson hires a consultant to develop a first draft of the standards.

A working group of 13 people (including a forest ecologist, forest economist, cultural anthropologist, sawmill operator, logger, forest manager, and landowner) is convened to review the draft.

Sigurd Olson develops a second draft based on the working group input.

1995   The revised draft is sent to over 100 people for comment, with 30 respondents.

The draft is revised again according to recommendations.

1996   A public forum is planned for 1996 to discuss the standards and elect an advisory body to review and formally endorse the standards.

Key Issues

The original impetus for the consultative process was the lack of standards in the region. This situation had become urgent, since certification assessments had already begun in the area. Therefore, the initial emphasis was structural—to quickly develop a set of standards with input from a small working group.

However, it is clear that the Sigurd Olson Institute also placed great importance on the symbolic nature of a consultative process—of maintaining transparency and encouraging broad public participation. This was evident by the large number of people consulted and the plans for a future public forum.

Swedish Consultative Process

WWF Sweden conducted several certification-related consultative processes. One of these was for the development of forest management standards for certification purposes in Sweden's boreal forests.

Process

1993   Initial interest generated by a consultative study conducted by WWF Sweden on the feasibility of certification in Sweden.

      WWF Sweden forms a "reference group" of various interest groups, including industry, forest owners associations, scientists, forest authorities, and environmental NGOs.

      Several scientific groups (primarily biodiversity experts) are convened to develop a technical draft.

      Two pilot studies are undertaken by the industry on tracking the chain of custody.

      Forest industries and forest owners associations publicly support the WWF-led process.

1994   The Swedish Society for Nature Conservation joins the project as a formal partner to WWF Sweden.

1995   A multistakeholder, independent, working group formed to review, revise, and formally adopt the standards.

Key Issues

The most important aspect of this consultative process was the emphasis on a structural approach. Emphasis was placed on drafting the standards—identifying and filling information gaps. In addition, the function of the reference group was primarily to ensure transparency and to ensure a mechanism for information sharing among key interest groups. Given the explicit support of

the various interest groups, a political approach would likely have been unnecessary.

Also emphasized in this process was the human resource approach. Reference group participants were not appointed to represent various interest
groups—they were selected on the basis of a sincere commitment to sustainable forest management.

FSC Consultative Process

The Forest Stewardship Council (FSC) designed and implemented a broad
consultative process to develop principles and criteria of forest stewardship
worldwide. These principles serve as a framework for the development of national and regional forest management standards.

Process

1992   FSC consultative group formed as a project of WARP (Woodworkers
        Alliance for Rainforest Protection, now known as Good Wood Al-
        liance).

        FSC consultative group prepares first three drafts of comprehensive
        forest management principles.

        Draft of principles is circulated to over 1000 stakeholders worldwide
        for comment.

        A meeting of 52 representatives from various stakeholder groups con-
        vened to review principles. Participants call for a broader consultative
        process.

        Working group established to refine document.

1992–   FSC working group develops two more iterations of the principles and
1993   circulates them to over 5000 stakeholders worldwide. FSC also con-
        ducts 10 in-depth country assessments, each of which reviews the for-
        est management principles.

1993   Two more drafts produced, based on input from consultative process.
        Each draft is circulated for review by colleagues of working group
        members and other interested stakeholders.

        FSC Founding Assembly convened, with 120 participants from 30
        countries. FSC Principles and Criteria accepted as a draft. Further
        consultations on specific areas requested by the membership.

1994   Second working group formed to focus on specific issues. New Prin-
        ciples and Criteria draft developed and circulated for comment. Ninth
        and final version submitted and formally ratified by membership in

1994. Plantation Principle submitted for further consultation and ratified in 1996.

Key Issues

The primary framework for the FSC's consultative process was political. The organization went to great lengths to allow multiple iterations of draft documents and adequate time for these drafts to be reviewed by various interest groups. Meetings and working groups were often seen as arenas for gaining influence in the certification process and for developing and maintaining positions vis-à-vis other interest groups.

The FSC was also deeply committed to some symbolic aspects of the consultative process. As an organization attempting to operate under a new sustainable development paradigm, it was held to strict democratic and participatory principles. Meetings operated by consensus, and when votes were formally cast, whether in meetings, working groups, or by postal ballots, they often reflected near unanimity.

Elements of a Consultative Process

The success of a consultative process in meeting multiple expectations and accommodating conflicting values and beliefs rests on careful planning and sensitive implementation. The following elements may prove useful to program managers attempting to implement a consultative process.

Blueprint Approach versus Learning-Process Approach

There are numerous models for designing problem-solving processes. These models nearly always include a linear succession of the following stages: (1) problem identification and definition, (2) project design, (3) project implementation, (4) project monitoring, and (5) evaluation and follow-up. This model, often called a "blueprint approach," relies on the widespread application of lessons learned from a particular pilot project, "much as a building contractor would follow construction blueprints, specifications and schedules" (Korten 1984).

An alternative to the blueprint approach is the learning-process approach, which begins with the assumption that initial conditions, problems, and solutions may be only partly understood at the outset of the process. A learning-process approach depends on the continual incorporation of new information into the identification of the problem, and the design and implementation of the process, in what has been called "a circular, progressively self-modifying loop" (Nolan 1988).

One difference between the two approaches lies in the role of the participants in the design and implementation of the process. While the degree to which participants control the process will vary, in general a blueprint approach assumes that a central authority is in control of the process. Friedmann (1984) stated that "Planners, whose background places them in an intellectual tradition that conceives of planning as physical design writ large, . . . conceive of planning, like any other activity, as being subject to the division of labor and, in this division, they mean to be the artisans of plans." A learning-process approach, however, depends on the participation of all stakeholders to shape the overall direction of the process and to assist at all design and implementation stages (Nolan 1988).

The Canadian Round Tables on the Environment and Economy include the learning-process approach as an important component of a successful consultative process (Anon. 1993). Participants should be encouraged to define the important issues, to clarify roles and responsibilities, and to establish ground rules. They add that the consultative process should be flexible in order to incorporate new learning into the process.

Appropriate Scale

One of the challenges to designing and implementing a consultative process is in determining an appropriate scale. In general, the scale of the consultative process should match the scale of the problem or project. An international policy level consultative process should involve international level stakeholders, while a regional or local consultative process should involve field managers, local NGOs, small businesses, and other local stakeholders within the region. That is not to say that local forest managers, for example, should be excluded from an international level consultative process. The challenge to designers of a consultative process is to incorporate mechanisms for eliciting a wide range of scale-specific information, and to coordinate this process to ensure that the end result is relevant and appropriate to the scale of the problem.

Consultative Process Techniques

There is no one best way to design or implement a consultative process. Indeed, a learning-process approach assumes that the technique and design of the process will be adapted to suit the constantly changing focus of the problem. However, a variety of information-gathering techniques, borrowed primarily from the social sciences, can be useful tools in the consultative process.

Some of these techniques include in-depth interviews, surveys, focus groups, peer review of draft documents, working groups, advisory councils, discussion papers, and public meetings. An important consideration in choosing an appropriate technique is the scale of the consultative process. In general, broad tools such as surveys and discussion papers are well suited for broader consultative processes, while narrower tools such as focus group studies and one-on-one interviews are well suited for local consultative processes. Another important consideration is the objective of the process. From a structural framework, information-gathering tools such as surveys and interviews will be important. From a human resource framework, consensus-building techniques such as focus groups, small meetings, and iterative peer reviews of technical drafts could be used. From a political framework, negotiating forums such as public meetings are appropriate. And from a symbolic framework, a highly visible and largely symbolic technique such as a discussion paper may be used.

Working Group

One of the more useful techniques in a consultative process is a working group. Working groups are one of the best ways for ensuring a balanced and representative decision-making process, for gathering and synthesizing information, and for reaching consensus on the conclusions. Working groups can easily be interpreted as an important aspect of the consultative process from all four of the organizational frameworks. As such, it is important to incorporate elements from each framework into the design of the working group.

From a structural approach, working groups should be designed to have both direct participation, in the form of voluntary, interested individuals expressing their own opinions, and representative participation, in the form of individuals representing an interest group, organization, or constituency. The working group should be well structured to ensure that new information from the consultative process is considered and synthesized.

From a human resource approach, participants of working groups should be carefully selected to ensure effective group dynamics. Individuals who are overly aggressive, dominant, reticent, or abusive may not be effective group members, even though they are highly experienced, articulate, well known, and represent a large interest group.

From a political approach, the working group should be designed to allow multiple iterations of any decisions or documents agreed by the working group, in order to ensure that coalitions which do not participate in the working group itself will have the opportunity to provide input.

From a symbolic approach, the working group, as a microcosm of democratic decision making, should adhere to democratic principles. In addition, the outcome of a working group process should be respected and adhered to in the implementation of the findings (Anon. 1993).

Continued Input

Regardless of the specific tools and techniques used in the consultative process, it is necessary, for structural, human resource, political, and symbolic reasons, that the overall design maintains a mechanism for continued stakeholder input.

Continued input and feedback are important from a structural perspective, as they allow the design and implementation of the process to emerge as a learning process rather than remain a blueprint approach. Such mechanisms should be able to incorporate new information on both the content of the process (such as principles, standards, and criteria for forest management) as well as the process itself (such as how the standards are developed and approved). In addition, input mechanisms should also provide for future input, as the application of the results of the consultative process will inevitably generate new learning. This is particularly important in the case of forest management standards, where the results are subject to periodic review and revision.

From a human resource perspective, mechanisms for continued input can be useful in continually tapping into the human energy that can be generated from a consultative process. This mechanism can be useful for establishing and maintaining networks and communications linkages among various individuals and organizations. Such mechanisms are important politically for their ability to allow coalitions to have continued access to influence the content, design, and implementation of the consultative process. This is particularly important when stakeholders represent larger interest groups and must conduct consultative processes of their own to develop a consensus or majority opinion among their constituencies. In addition, organizational philosophies can change over time, in a speciation of ideas according to internal and external organizational pressures. Furthermore, a mechanism for continued input, such as a monitoring committee or permanent consultative forum, can be an important tool for maintaining accountability to a variety of stakeholders.

From a symbolic perspective, continued input mechanisms are important as a symbolic exercise in participatory and democratic decision making. The Canadian Round Table Discussion cites a postagreement mechanism for input as one of the 10 principle elements of a consultative, consensus-building process (Anon. 1993).

Participation and Representation

Participation and representation are two of the most important aspects of a successful consultative process. These two concepts, which embody separate political philosophies, each have their own strengths and weaknesses. In an appropriate balance, a combination of the two can strengthen the overall consultative process.

Participatory decision making is a system in which stakeholders directly engage in expressing their own views and opinions. Strengths of this approach, in which "all parties with a significant interest in these issues [are] involved in the consensus process" (Anon. 1993), include a greater commitment to the process, as well as a greater depth and breadth of input. From a symbolic perspective, it is important that the process remain open and accessible to any interested party.

Representative decision making is a system in which participants primarily express the views and opinions of their constituencies (Swift et al. 1994). This approach has several strengths. "The concept of representation has both symbolic and substantive purposes—symbolic in that it demonstrates credibility and commitment to democracy, substantive in that it enhances the influence of users and ensures that the right questions are asked" (Swift et al. 1994).

In this approach, representatives are accountable both to their constituencies, as well as to the other participants involved in the process (Anon. 1993). One element to consider is the selection process of the representatives. While individuals may "represent" an interest group, such as an anthropologist representing a local community, it is important that interest groups have the opportunity to determine their own representatives, according to their own mechanisms and decision-making procedures, in order to maintain the integrity and credibility of a consultative process (Anon. 1993).

Regardless of the overall balance between representative and direct participation in a working group, the working group structure can be complemented by parallel information-gathering and decision-making techniques. Supplementing the working group with discussion papers, peer review of draft documents, and open public or electronic conferences ensures that both styles of decision making are utilized.

Incorporating Diversity

Fundamental to the nature of a consultative process is the successful integration of a diversity of ideas, opinions, perspectives, and worldviews. First and foremost in this diversity are the individuals involved. Each individual brings a unique personality, style, and voice to the consultative process. An extension of individual diversity is the range of agendas and objectives individuals have about the consultative process. Whether from a structural, human resource, political, or symbolic framework, each person interprets the purpose of the consultative process in a unique way. This diversity of individuals should be recognized, respected, and encouraged, in order to create a dynamic learning process.

Another important element to consider, especially in a broad consultative process, is the diversity of cultures that may be represented. Cultural differences in decision-making procedures, use and interpretation of silence, and facilitation and leadership styles can strengthen or weaken a consultative process, depending on the level of understanding of these differences among the participants.

Finally, a consultative process at any level will need to incorporate various scale-specific perspectives. With the FSC's consultative process, for example, incorporating a diversity of perspectives in scale meant bringing together managers of forest lands with forest policy makers, owners of small businesses with national industry trade organizations, and local grassroots NGOs with international NGOs.

Addressing Equity Issues

Chambers (1983) describes two cultures of development in his discussion of people-centered development. The first is the "insider culture," which consists of the poor, the rural, the agricultural, and the low status. The second is the "outsider culture," which consists of the rich, the urban, the industrialized, and the high status (Chambers, p. 4). In an international consultative process, this distinction can be broadened to include inequities between "southern hemisphere" and "northern hemisphere," or low income and high income countries.

For a consultative process to be just and inclusive, the design must contain mechanisms for addressing these and other traditional inequities—inequities of access to resources, to information, and to decision-making power. Indeed, the Canadian Round Table on Environment and Economy cites the commitment to providing equal opportunity to resources, information, and decision making as one of the key principles in a consensus-building process (Anon. 1993).

Maintaining Transparency and Accountability

An important element of a consultative process, particularly from a political and symbolic framework, is the maintenance of transparent and accountable processes. Two aspects of a consultative process can often fall under scrutiny: decision-making processes and financial accounting. In designing and implementing transparent decision-making procedures, it is important to ensure that decisions are clearly communicated and available to any interested parties. Decision-making processes themselves should be available for review. Providing information on how decisions are made, who has ultimate decision-making authority, and what mechanisms exist for future input or redress, are all steps to improve the transparency of the decision-making process.

Full and accurate accounting of both income and allocation of financial resources can also help to maintain accountability. Financial integrity and independence can be especially important when there is a potential conflict of interest between the source of funds and the outcome of the consultative process.

Concerns about accountability in a consultative process exist both internally and externally. Participants involved in a consultative process are accountable to other participants in the process, including other working group members, broader advisory groups, and, when applicable, a larger membership. Participants are also accountable to those not directly involved in the consultative process, including any constituencies they may represent, as well as the general public.

Some mechanisms for enhancing transparency and accountability include clear and detailed records of decisions and decision-making processes (such as minutes of meetings and working groups), accessible and efficient grievance procedures, and iterative decision-making procedures that allow for broader public participation.

Deciding by Consensus

A consultative process that involves a wide diversity of individuals, cultures, professions, and perspectives demands a decision-making style that is able to acknowledge and incorporate those differences. The style best suited to accommodating such diversity is consensus.

While an important element of consensus decision making is the outcome of a general agreement or accord, equally important is the process by which this accord is reached. A consensus decision-making process provides opportunities for participants to better understand each other by focusing on the underlying interests and needs of each participant, rather than focusing on previously defined positions or platforms (Anon. 1993; Fisher and Brown 1989). Furthermore, consensus decision making fosters greater commitment to the process and ensures that all participants have a stake in the final outcome.

Shared Ownership of the Process

One of the most controversial and problematic elements of a consultative process, particularly in a political framework, is the question of ownership of the process. Inevitably participants of a consultative process will raise the question: "Who controls the decision making at the end of the day?" The success of the consultative process may depend on whether this issue is fully and appropriately addressed in the planning and implementation stages. Problems in ownership of the process are likely to occur if the consultative process is interpreted from two different frameworks simultaneously (for example, structural and political). Ground rules should be established early in the process.

From a structural perspective, it is often clear that the designer and implementer control the decision making, since they are often the ones identifying the need for the information and providing the resources for the process itself. In such a process, participants may even have no desire to be involved in future decision-making processes.

From a human resources perspective, often the decision accurately reflects the consensus of the participants and is readily adopted. The final decision-making processes are likely to be only marginally important to the participants.

From a political perspective, control over the ownership is fundamental to the credibility and integrity of the consultative process; the final decision-making procedures and mechanisms are often part of the negotiations themselves. Participants may insist, for example, that the decision making is not owned by any one particular interest group, but that it is vested in an elected, multistakeholder group that is held accountable for its decisions.

From a symbolic perspective, it is important that grievance procedure mechanisms are established. These ensure that all decisions are, through due process, accessible to all possible stakeholders.

Limitations of a Consultative Process

Every consultative process will have some strengths and weaknesses. Each of the frameworks described in this chapter also have inherent strengths and weaknesses. The following are some of the limitations that may hinder any single approach or method of a consultative process.

Structural Limitations—Learning-Process Approach

An openness to change, uncertainty, and ambiguity characterizes the learning-process approach. While this model is an important element of a consultative process, there are limitations to its application. One of the difficulties in "learning as you go" is the relative inefficiency of the process. Part of the objective of a learning process is "learning to be effective, learning to be efficient, and learning to expand" (Korten 1984). When a consultative process uses the learning process as its model, the process may require an initially high input of resources for a relatively low return of information or results. Furthermore, this process can be time consuming, as more and more participants become involved in defining and designing the process. A time-consuming consultative process with an ever-changing problem definition must be balanced with a realistic time frame to bring focus to the process, conserve resources, and measure progress (Anon. 1993).

Human Resource Limitations—Representation and Participation

Involving a large number of participants in decision making, even at a local level, is "one of the fundamental constraints of participatory democracy" (Swift et al. 1994). Direct participatory approaches may not allow full, adequate, or accurate representation of diverse interest groups and may favor those who have the financial resources to participate. From a political perspective, direct participation may diffuse the strength of a coalition's negotiated position or platform, since each person's opinion is weighed equally and does not reflect the opinions of a constituency. The approach is also hindered by practical constraints such as transportation and infrastructural support, as well as by skepticism, disinterest, and participants' lack of confidence in the consultative process (Swift et al. 1994).

Representative decision making is not without its weaknesses, however. Often it is neither practical nor desirable to have an entirely representative system. Swift et al. (1994) write "the practice of representation is rather more complex . . . as questions of direct versus indirect representation are married with the identification of the right individuals and groups to be represented." The design of the consultative process should be flexible to include both direct and indirect representation at meetings and other decision-making arenas.

Political Limitations—Determining Appropriate Scale

Appropriate scale is also an important design issue, particularly at an international or national level when the numbers of potential stakeholders and participants far outweighs the logistical support to accommodate them all. To effectively address multiple scales at one time, working group participants should include as many varied scale-specific perspectives as possible.

A consultative process that addresses complex, multidisciplinary issues will necessarily involve a diversity of disciplines and backgrounds. The process of developing forest management standards for certification purposes, for example, requires participation from landowners, foresters, biologists, retailers, loggers, environmental NGOs, community groups, government representatives, and certifiers, at the very least. For a consultative process to successfully synthesize these differences, the entire group involved in the process must respect the differences inherent in a multidisciplinary group. An inherent limitation to a political approach can be the eventual collapse of negotiations when the process becomes extremely polarized or participants focus on positions rather than interests.

Symbolic Limitations—Finding Consensus

While consensus may be the most appropriate decision-making style for most consultative processes, there are limits to its application. The higher the number of participants, the more difficult consensus can be to achieve, especially if the forum is not conducive to small-group discussions. Consensus decision making may be appropriate at some stages, such as the development of a draft, or other interim stages, whereas modified consensus or other styles may be more appropriate in the final ratification stages.

Relevance to Future Certification Initiatives

Two tools are presented below for the designers and implementers of a consultative process: a matrix for analyzing the various elements and framework approach and a set of questions to consider.

Matrix for Analysis

The matrix (Table 2.2) summarizes the main elements of the four framework approaches to a consultative process. While all of these elements may be present at any one time, designers and implementers of a consultative process

Table 2.2
Matrix for analysis

Framework	Structural Approach	Human Resources Approach	Political Approach	Symbolic Approach
Objectives	To gather information and to understand the problem	To build consensus, to establish networks, to release social energy	To negotiate positions, to gain power, to influence outcomes	To create meaning through metaphors and symbols
Key Issues	Learning-process approach; appropriate scale	Managing diversity; adequate participation	Equity of access to resources, information and decision making; authentic representation	Appropriate decision-making style (consensus driven); transparency; shared ownership
Techniques	Surveys, interviews, rapid rural appraisal	Focus groups, small meetings	Public meetings, workshops, and conferences; peer review of draft documents	Discussion papers
Working Group Issues	Balance of direct and representative participation	Appropriate individual for effective group dynamics	Multiple iterations of working group decisions to allow constituency consultations	Working group adheres to democratic principles
Continued Input	A mechanism to facilitate the learning process	A mechanism for tapping into human resources; for building networks and strengthening relationships	A mechanism for maintaining continued access to the process by various interest groups	A mechanism that embodies transparency and participatory democracy

may find it useful to reflect on the overall approach to be taken, according to local needs, circumstances, and resources, and to design and implement the process accordingly.

Questions to Consider
When Designing a Consultative Process

The following questions are intended for designers and coordinators of a consultative process. These questions may help to ensure that elements from each of the various frameworks are included in the process, and that the process is appropriately modified to local circumstances.

• Framework: From what framework—structural, human resource, political, or symbolic—is the consultative process likely to be interpreted? Which framework will be dominant? How will the design and implementation of the process incorporate multiple frameworks in order to satisfy multiple expectations?

• Design: What mechanisms exist in the design process to ensure a learning-process approach instead of a blueprint approach? How will the design deal with ambiguity and uncertainty? To what extent are the participants involved in the design phase? What mechanisms exist for adapting the design to accommodate new learning?

• Scale: Does the scale of the consultative process match the scale of the problem? Are there mechanisms for enriching the diversity of participants by including individuals with smaller and larger scale perspectives?

• Techniques: Are the techniques of the consultative process appropriate to the particular framework? If a working group is included in the process, is it structured to accommodate the needs of each framework? What mechanisms and structures exist for complementing the working group?

• Continued Input: Does the consultative process have clear mechanisms for continued input into both decision making and the overall design and implementation of the process? What mechanisms will exist in the future?

• Participation and Representation: How will the consultative process strike a balance between direct and representative decision making? What mechanisms exist to enhance aspects of each style, in order to strengthen the overall consultative process? Who decides or limits who participates in any phase of the process? Are there grievance procedures for those who feel excluded from the process?

• Incorporating Diversity: What is the extent of individual, cultural, and professional diversity in the consultative process? How does the process build on these differences to create new knowledge and new perspectives?

• Equity: How does the consultative process recognize and address traditional inequities in access to resources, to information, and to decision-making power? How will subsequent implementation or future initiatives continue to address these inequities?

• Transparency and Accountability: How will the consultative process develop and maintain mechanisms for ensuring transparency and accountability, both internally and externally? How are decisions and processes communicated, and what avenues exist for redress?

Conclusions

Critical to the success of any certification program is the extent to which affected stakeholders can agree on local forest management standards. A broad consultative process can facilitate the development and acceptance of these standards.

There are various issues to consider when designing and implementing a consultative process, including which key elements to consider in the process and which frameworks for analyzing the consultative process from various perspectives should be used.

References

Anon. 1993. *Building Consensus for a Sustainable Future: Guiding Principles.* Round Tables on the Environment and Economy in Canada. Toronto, Canada.

Bolman, Lee G., and Terrence E. Deal. 1990 *Modern Approaches to Understanding and Managing Organizations.* San Francisco: Jossey-Bass.

Chambers, Robert. 1983. *Rural Development: Putting the Last First.* New York: John Wiley & Sons.

Dodge, William, ed. 1978. "Consultation and Consensus: A New Era in Policy Formulation?" Report from the Compensation Research Center of The Conference Board in Canada, December 1978.

Fisher, Roger, and Scott Brown. 1989. *Getting Together: Building Relationships as We Negotiate.* New York: Penguin Books.

Friedmann, John. 1984. "Planning as Social Learning," in *People-Centered Development.* West Hartford, CT: Kumarian Press.

IUCN, UNEP, and WWF. 1991. *Caring for the Earth: A Strategy for Sustainable Living.* Gland, Switzerland.

Korten, David C. 1984. "Rural Development Programming: The Learning Process Approach," in *People-Centered Development.* West Hartford, CT: Kumarian Press.

Korten, David C., and Rudi Klauss, eds. 1984. *People-Centered Development: Contributions toward Theory and Planning Frameworks*. West Hartford, CT: Kumarian Press.

Nolan, Riall. 1988. *Project Design: A Problem-Solving Approach to Planning Community Change*. Brattleboro, Vermont: School for International Training.

Sizer, Nigel. 1994. "Opportunities to Save and Sustainably Use the World's Forests through International Cooperation." World Resources Institute Issues and Ideas Report, December 1994.

Swift, Paul, Gordon Grant, and Morag McGrath. 1994. *Participation in the Social Security System: Experiments in Local Consultation*. Hants, England: Avebury Press.

Chapter 3

The Development of Standards

Jamison Ervin and Chris Elliott

Standards, which are "an acknowledged measure of comparison for quantitative or qualitative value" (American Heritage Dictionary 1980), exist for a wide range of products and processes ranging from camera films to quality control. In forest management certification, standards are the primary tool by which an independent assessor evaluates forest management practices. In this chapter, standards are defined as a measure for comparing existing management practices within a particular forest unit against a set of ideal principles or conditions. Standards that address ecological, social, and economic issues do not directly measure the sustainability of forest management. This would require a full knowledge of the long-term impacts of management activities on forest health—knowledge we do not currently, nor may ever, possess. Rather, standards measure the application of acceptable forest management practices for a given area.

While standards are the basis of any certification assessment, they are not in themselves sufficient for a credible certification program. Of equal importance are the process by which the standards are developed and adopted, and the system for using the standards in a forest evaluation (evaluation protocol). These topics are addressed in Chapters 4 and 5, respectively.

Types of Standards

There are various types of forest management standards. Some of the differences among these include the issues addressed, the focus of the evaluation, the intended use, the scale of the forest unit, and the breadth and depth of the consultative process. The following discussion explores some of these differences and examines which standards are currently being used for certification purposes.

Basis for Evaluation: Performance-Based versus Systems-Based

Performance-based standards are intended to be used to evaluate how well a forest manager performs or follows the best management practices. This type of evaluation typically includes performance measures in such areas as appropriate silvicultural techniques, maintenance of biological diversity, local rights and benefits, and economic viability of the operation. Nearly all forest management standards include ecological, social, and economic issues. There are, however, some differences in the relative emphasis placed on each of these aspects. (See Chapter 5 for further discussion of the certification assessment.)

Systems-based standards are based on the assumption that if a company or forest manager has an adequate system in place for dealing with the environmental impacts of its activities, this will provide a sound basis for minimizing negative environmental impacts. Certification that a company has an environmental management system in place does not guarantee that certain levels of performance are achieved, however. Instead, a certifier evaluates how well a forest manager incorporates environmental objectives and targets into an overall management system, and how well that system is being implemented. Performance levels may be set by the company or operation but are not part of the standard. This approach, which has been largely developed through the International Organization for Standardization (ISO) 9000 series of standards on quality management, is now being applied to environmental management through ISO's new set of standards, the ISO 14000 series. An example from the British Standards Institute (BS) of some of the requirements of system-based environmental standards are listed in Table 3.1.

Most forest certification programs are based on performance standards, although the Canadian Standards Association has been working on a set of system-based standards for forest management in Canada. These were submitted as a proposal to the ISO in 1995 as a potential basis for an International Standard in the ISO 14000 series. This proposal was withdrawn in 1995 partly because many environmental groups objected to the process by which the standards were being drafted, and also because of the lack of focus

Table 3.1

Requirements for environmental management systems

BS 7750's main requirements for a company to be certified as having an environmental management system:

1. A documented environmental management system that takes account of any pertinent code of practice to which it subscribes
2. A documented environmental policy
3. Clear definition of all organizational responsibilities
4. Procedures to ensure that contractors are made aware of all environmental requirements
5. Procedures for receiving and responding to stakeholder views
6. Procedures to specify environmental objectives and relevant targets
7. Establishment and maintenance of a program's plan, with the objectives and targets
8. A documented system including an environmental assessment manual
9. Procedures which ensure that activities with a significant environmental impact take place under controlled conditions
10. Procedures for verification of compliance with specified requirements, including targets
11. Procedures for initiating investigation and corrective action in the event of noncompliance
12. Records which demonstrate performance of the environmental management systems

---

on forest management practices on the ground. The proposal is currently being debated (as of summer 1996) in technical ISO working groups and may be resubmitted to the ISO for consideration in 1997.

These two types of standards are distinct in theory; in practice they often overlap. For example, the FSC, which is primarily performance-based, includes elements of systems-based standards in its accreditation program. Certifiers must implement total quality management systems (FSC 1995), and the Principles and Criteria require periodic monitoring of management plans and forest conditions (FSC 1994). Eventually certifiers may incorporate elements from both performance-based and systems-based standards into their evaluation systems.

Implementation

Forest management standards are developed for different purposes. They may be developed as a set of voluntary standards, or they may be intended for use in national laws and international agreements.

Throughout this book, certification is understood to include both performance and systems-based approaches, but only when these standards are applied voluntarily. Mandatory implementation of standards is considered a requirement in law and therefore part of regulatory enforcement, not certification.

Voluntary standards usually describe best management practices and ideal forest conditions. They are developed as a set of attainable but rigorous goals to which forest managers, loggers, and forest owners may subscribe. Such standards may be used as part of a self-assessment guide, such as the "best management practices" for logging common throughout the United States. (See Box 3.1.) Or they may be used as part of an independently verified and monitored system, such as independent certification.

Mandatory regulations usually describe practical approaches and easily obtainable forest conditions. They are developed as the minimum legal requirements to which all forest managers, loggers, and forest owners must subscribe. These regulations or requirements are written into laws as part of a forestry act or a code of forestry practices.

Some voluntary standards may be general, such as the American Forest and Paper Association's Sustainable Forestry Principles, and some mandatory regulations are quite specific, such as British Columbia's Forest Practices Code. Usually voluntary standards aim at the highest common denominator in forest management, and mandatory regulations at the lowest.

Standards can be further differentiated by examining the scale of the forest unit, and the process by which the standards are developed.

Box 3.1

Best management practices

In the early 1990s, the U.S. government passed legislation requiring that each state develop voluntary best management practices to help control non–point source pollution. All 50 states now have voluntary non–point source pollution practices that provide farmers, foresters, and other landowners with guidance on how to control erosion. The "best management practices" were developed after consultation with scientists, landowners, foresters, environmental groups, and industry. A few states have begun to develop similar practices for other aspects of forestry, such as silviculture practices, wetlands, and biological conservation. Such practices provide a useful performance-based starting point for certification standards.

Differences in Scale

Forest management standards as defined in this chapter apply to the individual forest management unit. Global forest guidelines and principles are intended to apply to all forests. One example of such principles are the Forest Stewardship Council's principles and criteria of forest management, which apply to all types of forests worldwide.

Principles, guidelines, or standards may also apply to a region, which may be defined by forest type, biophysical characteristics, or political boundaries. Principles apply to a general forest type, such as temperate and boreal forest management—for example, the Santiago Declaration (Canadian Forest Service 1995); standards apply to a narrowly defined forest type, such as the management of northern mixed hardwoods—for example, Lake States Standards (Sigurd Olson Environmental Institute 1994).

Biophysical characteristics such as mountain ranges or river basins may also define the parameters of forest management guidelines, such as the Amazon Pact Agreement for forest management in the Amazon region. Political boundaries are perhaps the most common demarcation of standards. Examples include standards for Sweden, Indonesia, New England, the United Kingdom, and the Pacific Northwest.

Standards may also exist at a very local scale. Individual forest management units may have a distinct set of standards developed specifically by the forest landowner, manager, or certifier. These are often referred to as an evaluation protocol and may include detailed, specific procedures that managers need to follow in order to meet the regional standards.

The large differences in perspectives in scale do not necessarily lead to incompatible perspectives among principles, criteria, standards, and protocol at the global, regional, and local levels; they are in fact complementary. The following example of principle, criterion, and standard illustrates how global principles and local standards interrelate.

Principle

Forest management shall conserve biological diversity and its associated values, water resources, soils, and unique and fragile ecosytems and landscapes, and, by so doing, maintain the ecological functions and the integrity of the forest (Principle #6 in FSC Principles and Criteria).

Criterion

Ecological functions and values shall be maintained, enhanced, or restored, including (a) forest regeneration and succession; (b) genetic, species, and ecosystem diversity; and (c) natural cycles that affect the productivity of the forest ecosystem (FSC Criterion; 6.3).

Standard

No activities are permitted in the following habitats (except when needed to sustain natural biological diversity): . . . nonproductive areas with less than one cubic meter annual production per hectare, wetland forests, and undrained, low-productivity peat lands (Swedish draft standards; 1.c).

Management or Evaluation Protocol

Current maps of the forest management unit show where all such nonproductive areas exist. Information about the importance of these areas is included in training programs. Such areas are regularly monitored by the management for compliance.

One final aspect about scale is the focus of certification itself. Most certification programs focus on the forest management unit—that is, the area for which a forest manager is responsible. While some programs may consider national-level certification, this chapter focuses on certification that specifically targets individual forest management units.

Breadth and Depth of Consultative Process

Another area in which standards or principles may differ is the extent to which they are submitted to a consultative process. Some standards and principles are developed in a multistakeholder consultative process, which includes large numbers of stakeholders from diverse backgrounds. Examples of standards and principles developed and approved by a broad consultative process include the Swedish standards, the Lake States standards, and the FSC Principles and Criteria. (See Chapter 2 for a description of these processes.)

Standards may also be developed with little consultation, or consultation with a relatively narrow stakeholder focus. Sector-specific standards include industry associations such as the American Forest and Paper Association's sustainable forestry guidelines, NGOs such as Greenpeace's clearcut free standards, and standards-writing bodies such as the American Standards and Testing Materials (ASTM). Sector-specific standards typically focus on consultation within a specific constituency or community of interests.

Purpose of Standards

Standards are an integral part of the certification process because they serve as the basis for assessing the quality of the forest management. Standards allow certification to maintain objectivity and provide the basis for appeals.

Standards are intended to be clear, easily applied, and are unbiased toward any one particular interest group. Standards are designed to ensure that an assessment of one forest management unit by two different evaluators, or even two different certifiers, will yield similar results. While certifiers may have different protocol for interpreting and evaluating standards, and while any assessor must use professional judgment in applying standards, consistent standards foster overall objectivity and consistency in the certification process.

Standards are also a useful mechanism for ensuring the transparency of certification. If a local community wishes to dispute a certification decision, the standards can provide an objective basis for disputing the certifiability of the forest operation. Standards also provide the basis for any legal claims if a retailer of a certified product is sued for false advertisement.

Standards Development—Key Players

Standards can be developed by any number of key stakeholders, including academic and research institutions, nongovernment organizations, forestry associations, certifiers, governments, and industry associations. Such standards in themselves are an insufficient basis for credible certification, however. To ensure that standards accurately reflect a wide group of stakeholder perspectives, any set of standards must undergo a consultative process. Both public consultation and an endorsement by an independent multisectoral group are required of standards in the FSC's accreditation process.

Key Elements

For standards to serve as a credible, objective, and effective basis for certification, they should have certain key elements. These include the following (adapted from LaPointe 1995):

- Credible to the public
- Supplier/consumer focused
- Single overall system
- International equivalence
- Compatible with relevant principles and criteria as well as with legislation

- Equitable for all users
- Practical in application
- Voluntary
- Auditable by a third party
- Continual improvement incorporated
- Accessible to small and medium enterprises
- Adaptable to different jurisdictions
- Adaptable to different ecological systems

Conclusions

The number of forest management standards has grown considerably over the past few years. These standards have been developed in various ways, by various interest groups, and for various purposes. The preceding parameters may be useful to those individuals and groups interested in developing standards for voluntary, independent certification systems. Performance-based and systems-based standards (or a combination of both) may be used in the evaluation.

While existing regulations will provide a useful starting point, standards should go beyond the legal requirements for the area, in order to avoid duplicating legislation. Adherence to the standards are intended to be independently and voluntarily assessed by certifiers. Global principles and criteria can provide a framework for developing regional standards; regional standards can provide a framework for developing specific evaluation protocol. Certification applies to individual forest management units, not large units such as entire countries or regions. Any group or individual may develop standards, but if they are to be a credible basis of certification they must be submitted to a broad consultative process.

Credible standards are the basis and the beginning of the certification process. The application of these standards in field-level assessments is the subject of Chapter 6.

References

Canadian Forest Service. 1995. "Criteria and Indicators for the Conservation and Sustainable Management of Temperate and Boreal Forests—The Santiago Declaration." Natural Resources Canada, 351 St. Joseph Boulevard, Hull Quebec.

FSC. 1994. "Principles and Criteria of Natural Forest Management." Forest Stewardship Council, Avenida Hidalgo 502, Oaxaca 68,000, Oaxaca, Mexico.

FSC. 1995. *Manual for Evaluation and Accreditation of Certification Bodies.* Oaxaca, Mexico.

LaPointe, Gerald. 1995. "Sustainable Forestry Certification." Unpublished paper.

SGS-Forestry. 1995. "Environmental Management Systems." Paper submitted to WWF Sweden seminar Forests for Life.

Sigurd Olson Environmental Institute. 1994. Lake States Standards.

Swedish Society for Nature Conservation. 1995. Draft Criteria. Stockholm, Sweden.

Chapter 4

Accreditation Process

Jamison Ervin, Chris Elliott, Bruce Cabarle, and Timothy Synnott

Accreditation, which is essentially a nongovernmental, and hence nonregulated, process (Pinkham 1952) can be defined as "a procedure by which an authoritative body recognizes the competence of a group to conduct . . . assessment activities" (Barrett 1993). It is a process of "certifying a certifier." When accreditation refers to forest management certification, it refers to the process of evaluating, endorsing, and monitoring organizations that independently conduct forest management assessments and chain-of-custody audits.

Such assessments may take many forms. These include countrywide certification, life-cycle analyses, report cards, environmental management appraisals, and criteria-based schemes (those programs that focus on the quality of the forest management practices within a particular forest management unit). Forest product certification schemes may also include first-, second-, and third-party assessments. First-party assessments involve an internal assessment of a company's systems and practices. Second-party assessment involves a second party, usually a direct customer of the product or a trade association, who assesses the company according to the customer's needs and any existing contractual obligations (Barrett 1993). Third-party assessment involves a credible and separate third party who assesses the forest operation based on existing agreed-upon standards. The third party then issues a "certification of conformity," and continues to monitor the certified operation (Barrett 1993).

This chapter primarily addresses the accreditation of independent, third-party, voluntary, and performance-based certification schemes.

In the United States, accreditation of certifiers, auditors, and assessors began during the early 1900s in what has been called "the age of standards" (Young et al. 1983). Some of the first independent accreditation bodies in the United States were established to develop common definitions and standards for colleges and preparatory schools. Within a decade, the "radically new concept" of accreditation had been adopted throughout the United States (Young et al. 1983).

Accreditation has never been a simple process. Fred Pinkham, the first executive director of the National Commission on Accreditation in the United States, described accreditation in 1952 as "an elusive nebulous jellyfish term that means different things to the same people . . . people who do not agree on what it is on which they do not agree and, I might add, on which they disagree violently, emotionally, and dogmatically."

Though still elusive and contentious, certification and accreditation have become more commonplace and accepted concepts, due largely to the internationalization of business; the increased relation between standards, quality, and competitiveness; and a public recognition of the need for harmonization of standards and certification procedures (Hutchins 1993).

Goals of Accreditation

The goals of accreditation are threefold: to provide consistency among certifiers and standards, to ensure the credibility of certification programs to the public, and to verify the integrity of a certifier's claims.

Consistency

Consistency is especially important when comparing one certification program with another. As mentioned earlier, certification may take many forms, including first-, second-, and third-party schemes, as well as life-cycle analyses, environmental management systems, report cards, and single-attribute evaluations. Accreditation helps consumers to differentiate between an internal assessment of the sustainability of one's own forest management practices and a third-party independent certificate based on externally recognized and accepted standards.

Consistency is also important in the marketplace. A recent study conducted by the Fauna and Flora Preservation Society found over 600 claims of sustainability on forest products marketed within the United Kingdom (Read 1991). These labels ranged from simple claims regarding the country of origin and tree planting, to more complex claims regarding management practices, use of plantation-grown timber, and other factors. On further investigation, only three were willing and able to substantiate their claims when challenged with a false advertising suit by the World Wide Fund for Nature in the United Kingdom. Accreditation provides clarity and consistency in marketplace claims, by laying ground rules for certification labels and by defining a single set of principles by which all forests, and forest product certifiers, will be evaluated.

For the timber industry, one of the most important aspects of accreditation is consistency within international trade. Some authors have directly equated international certification and accreditation with global competitiveness (Barrett 1993; Hutchins 1993). One of the direct benefits of international accreditation is not only agreement on a basic level of quality, but also assurance of access to a global economy in an "internationally recognized system of quality assurance" (Hutchins 1993).

Credibility

The second goal of accreditation is to ensure the credibility of certification claims to consumers and other key stakeholder groups. In the past, certification and accreditation have been driven by the ISO (International Organization for Standardization). ISO and national standards bodies (such as the American National Standards Institute in the United States) facilitated the development of standards *for and by industry and product groups* (Hutchins 1993). When the standards involved technical issues such as the speed of camera film, credibility was relatively easy to establish. Because the notion of forest stewardship inherently involves societal values, it requires a standards development process that involves far more stakeholders than simply industry and product groups. The narrowness of ISO's "community of interest" may be an inhibitive factor for it to succeed in the development of forest management standards.

Nonetheless some industry groups, such as the Canadian Pulp and Paper Association, have looked to the ISO for credibility in the development of standards, certification, and accreditation schemes. In Canada, the timber industry has promoted the development of standards through the Canadian Standards Association, the national member of the ISO. "To be credible," the industry states, "the standards must be developed by an independent, recognized body to guarantee both neutrality and objectivity." They add: "In Canada, the Canadian Standards Association is seen as the most appropriate organization for developing sustainable forestry standards" (CPPA 1995).

However, ISO member organizations may often have only marginal influence in approving standards. The American National Standards Institute, for example, screened or approved less than 25 percent of all nongovernmental standards developed in the United States. The rest were developed by nongovernmental bodies (Hutchins 1993).

Accreditation exists in a complex political, social, and economic context. By its standard-setting nature it affects a wide range of related organizations and individuals, which are often referred to as the "community of interest" for the accreditation body. Increasingly, accreditation has begun to account for the views and demands of this larger community of interest, particularly when dealing with environmental issues. The increase of "due process of accounting" for a wide range of stakeholder views may partly be attributed to the rapid advance of technology, a shift to a global economy, and the increased influence of nongovernment actors (Hutchins 1993). The importance of due process is likely to increase as these trends become more pronounced.

In particular, nongovernmental organizations (NGOs), especially those involved in forest conservation and community forestry issues, have played a key role in developing and promoting forest product certification. Groups such as the World Wide Fund for Nature, Greenpeace, and the National Wildlife Federation, among others, have been instrumental in providing support to the forest product certification and accreditation systems that account for the "NGO community of interests." These groups are therefore credible both to the NGOs themselves and to their broader constituencies. Similarly, progressive companies such as members of the 1995 group in the United Kingdom have been extremely influential in promoting certification within their own constituencies.

Ultimately consumers will determine the credibility of an accreditation system according to how well the system represents their particular values, as articulated by particular interest groups such as social and environmental NGOs.

Reliability

The third goal of accreditation is to ensure the reliability and integrity of the certifying organization. Such integrity includes efficient and reliable certification systems, organizations, and processes, including effective overall supervision, rigorous forest management standards, and demonstrably clear and valid public claims.

An evaluation of the organization's reliability essentially ensures that "the institution or program is what it says it is and does what it says it does" (Young et al. 1983). Such an evaluation also ensures that the organization is competent, independent, and transparent. Competency can be assessed by evaluating the certifier's organizational structure, recruitment, and training of personnel; internal management review policy; management systems; adherence to applicable laws; and systems for continuous improvement. Independence can be assessed by evaluating the financial means of support of the organization, the system for identifying and addressing potential and existing conflicts of interest, and by the certification application procedures. Transparency can be assessed by evaluating (1) the system for maintaining documents, general record-keeping practices, and public disclosure; and (2) the system for responding to complaints and public grievances (FSC 1995; ISO 1994).

An assessment of an organization's reliability also involves a review of the certifier's policies and procedures. This type of assessment ensures that the certifier is not only competent, but that it applies consistent and rigorous evaluation procedures. This part of an accreditation assessment looks for "conditions believed to be necessary and desirable," and to assess whether the program does, in fact, achieve quality (Young et al. 1983). Accreditation also evaluates the extent to which systems are in place for continual improvement. Reporting systems, peer review policies, and decision-making mechanisms are used to assess consistency, while policies for developing or adopting standards and methodologies for evaluating forest operations are used to assess organizational rigor and quality (FSC 1995; ISO 1994).

Effective overall supervision is critical to maintaining the reliability of a certification system. An accreditor plays a major role in providing this supervision, by evaluating and accrediting certifiers and by regularly monitoring the performance of the organization as well as the forest operations being certified. However, the integrity of an accreditor that is a member organization, such as the FSC, depends on the vigilance of its members in detecting and reporting infractions, and in pressing for continual improvements.

Finally, accreditation aims to ensure the validity, integrity, and reliability of a certifier's claims, and to maintain proper control over the use of certificates, logos, and labels (Cabarle et al. 1995). This control is maintained by a careful verification of the certifier's "chain-of-custody" monitoring procedures. (See also Chapter 6 for further discussion on the chain of custody.) Accreditation procedures may rely on a certifier's written methodologies, which must include annual and/or random inspections and clear documentation, separation, and demarcation of certified products, to determine whether a certifier is able to maintain control over the use of its certification label (FSC 1995).

This in turn provides assurance to consumers that a product with an accredited certification label does, in fact, come from a well-managed forest.

Key Elements of an Accreditation Process

Accreditation decisions are based on several elements. These include the standards used by the certifier in the evaluation; the certifier's competency, including the organizational structure and the evaluation procedures; and other key elements.

Standards

The most controversial aspect of any accreditation program is the process of defining standards. "Accreditation by its very nature," Pinkham (1952) claims, "represents a struggle over standards."

Accreditation also strives for the harmonization of standards, or the creation of global, uniform standards that apply universally (Hutchins 1993). This process can lead to a harmonization upward (seeking the highest common denominator) or a harmonization downward (seeking the lowest common denominator). For that reason, standards development, particularly in a field as polarized as forest management, can become highly controversial.

Nevertheless, a number of initiatives have begun to define forest management standards, for both certification and accreditation purposes. Ultimately, accreditation should ensure that the standards developed in, for example, Brazil will have the same level of integrity and international acceptability as standards developed in, for example, Sweden.

Similarly, accreditation should ensure that certified forest operations within the same region are judged according to the same set of standards. The two main questions for forest operations and, by extension, certifiers are "What are the standards?" and "Does the organization meet those standards?" (Young et al. 1983). Accreditation can clearly and definitively answer these two questions, based on standards that are explicit, agreed upon, and equally applied with consistent methodologies.

Competency

The competency of the certification organization is evaluated on the basis of the overall soundness of the organization and the integrity of the certification system.

An organizational assessment evaluates the competency of a certifier to

conduct forest evaluations and chain-of-custody audits based on the soundness of the organizational systems. Elements in this evaluation might include, for example: (a) clear organizational structures; (b) sufficient means of
financial support; (c) recruitment and training of qualified personnel; (d) adequate record keeping, especially of the chain-of-custody audits; (e) adequate
control over licenses and logos; (f) quality management systems reviews; and
(g) adherence to national laws (FSC 1995).

Certifiers must also have an explicit corporate policy regarding their commitment to total quality management and continual improvement, and a formal system in place to systematically monitor, review, and upgrade performance on a regular basis. Furthermore, certifiers must have formal systems
and processes in place that lend themselves to monitoring, review, and upgrading. This ensures that problems are identified and solved before they become liabilities.

Credibility

An assessment of the certification system also includes an evaluation of the
credibility of that certification system. Specific elements that would indicate
the credibility, reliability, and transparency of a certifier might include the
following: (a) a clear policy for identifying or developing forest management
standards, (b) transparent reporting procedures, (c) a peer review policy, (d)
clear certification decision-making procedures, (e) a process for suspending
and withdrawing certification if necessary, and (f) a fair and adequate grievance procedure and policy (FSC 1995).

Other Characteristics of Accreditation

To be credible and effective, an accreditation system should be universal, independent, voluntary, participatory, equitable, and transparent. If an accreditation system is to be recognized and provide market access worldwide, it
must be universal (Hutchins 1993). This includes the accreditation organization, its standards, and its evaluation procedures, all of which must be equally
and universally applied to any forest product certifier.

An accreditation system, as with any third-party certification scheme, must
be independent and not unduly subject to political, social, or economic pressures or other forms of influence (Hutchins 1993). In particular, a forest
management accreditation system must be independent from timber trade
influence. Maintaining an independent funding source and designing mechanisms for identifying and resolving conflicts of interest are two important
steps for ensuring an accreditor's independence.

An accreditation system is, by definition, a voluntary process (Young et al. 1983). As such, it can be a useful mechanism for regulating the timber trade. Such a process does not violate existing World Trade Organization agreements. Once certification and/or accreditation become mandatory, however, they may conflict with World Trade Organization agreements as a barrier to free trade (Droogsma et al. 1991).

Chapter 2 emphasized the importance of participation and due process in the certification and accreditation processes. An accreditation system must maintain fair and equitable avenues of participation and clear grievance procedures if it is to account for the views of a wide community of interests.

Similarly accreditation must be nondiscriminatory in the execution of its programs. Certifiers that are small, nonprofit, regionally based, and/or from developing countries must have equal access to the accreditation process as those from large, international, for-profit companies. Tailoring the fee and administrative requirements to suit the scale of the certifier can help to ensure equal access to the accreditation process.

Finally, the accreditation organization, the standards used, and the procedures followed must all be transparent and fully accountable to the public. This feature is the "hallmark of excellence" of accreditation bodies (Young et al. 1983). Transparency can be established by public disclosure of all relevant aspects of the accreditation body, as well as by clear and explicit decision-making organizational procedures.

Example of an Accreditor—Forest Stewardship Council

The Forest Stewardship Council (FSC) was established in 1993 as an independent, nonprofit accreditation body for certifiers of forest management systems and forest products. The FSC focuses solely on accrediting third-party, single-attribute, forest management unit certification programs and related chain-of-custody monitoring programs. To date it is the only operational accreditor of forest management certifiers. However, as forest product labeling becomes an accepted practice, new accreditation programs may be established, including one under the auspices of the ISO.

Process

The FSC's accreditation program involves a series of steps from an initial preapplication to the formal accreditation decision. The application steps are outlined below (FSC 1995):

- Preapplication and information gathering

- Formal application by the certifier, including a letter of intent, a description of the organization, certification procedures, chain-of-custody procedures, forest management standards, and a description of all certified forest operations
- Evaluation of the certifier's major documents
- Site visit and evaluation at the certifier's headquarters
- Site visit and evaluation of certified forest operations
- Formal accreditation decision
- Monitoring of the accredited certifier

Basis of FSC Evaluation

The FSC bases its evaluation of a certifier's organizational competency on adherence to guidelines for certifiers and principles and criteria of forest management. (See Table 4.1.)

The FSC also requires that local stakeholders are adequately consulted in the development of local standards, and that local or national FSC working groups (if they exist) endorse these standards. These additional requirements ensure the overall integrity and high quality of the forest management standards.

In addition to compliance with FSC principles and guidelines, the FSC requires certifiers to consult with any established FSC initiatives in the region of operation before certification decisions are made. The FSC itself must also consult with local FSC initiatives prior to making any accreditation decisions. These procedures ensure that while the certification and accreditation decisions are based on global principles, they are grounded in local circumstances and judged according to local sensibilities.

Conclusions

Nearly a century ago, Woodrow Wilson addressed the Middle States Association of Colleges and Schools by stating: "We are on the eve of a period when we are going to set up standards. We are on the eve of a period of synthesis when, tired of this dispersion and standardless analysis, we are going to put things together into something like a connected and thought-out scheme of endeavor" (Wilson 1907).

With a shift toward an increasingly global economy, a rise in the influence of citizens and nongovernmental organizations, and a growing recognition of our political, social, economic, and ecological interdependence, we may yet be "on the eve of a period when we are going to set up standards."

Table 4.1
Basis and elements of FSC accreditation

Basis of evaluation	Elements considered in an evaluation
FSC Principles • Compliance with laws and FSC principles • Tenure and use rights and responsibilities • Indigenous peoples' rights • Community relations and workers' rights • Benefits from the forest • Environmental impact • Management plan • Monitoring and assessment • Maintenance of natural forests • Plantations (FSC 1994a)	*Organizational Structure and System* • Description of organizational structure • Quality management system • Financial stability and resources • Independence • Recruitment and training of personnel • Documentation • Record keeping • Confidentiality • Control over certifications, marks, and claims • Reciprocal recognition of FSC accredited certification bodies • Public information • Complaints, appeals, and disputes • Management review • Communication with FSC • Compliance with national legislation (FSC 1995)
Guidelines for Certifiers • Compliance with FSC • Independence • Sound evaluation procedures • Transparency • Reciprocity among FSC-accredited certifiers • Public information • Verifiable chain of custody • Compliance with applicable laws • Equity of access • Maintenance of adequate documentation • Appeal procedures • Integrity of claims (FSC 1994b)	*Certification System* • Information for applicants • Information for assessors • Application procedure • Forest Stewardship Standards • Procedures carried out prior to assessment • Methodology for forest management evaluation • Methodology for chain-of-custody verification • Review of certification reports • Certification decisions • Appeals against certification decisions • Methodology for monitoring of certified operations • Maintenance and extension of certificate • Suspension and withdrawal of certificate • Changes to certification requirements (FSC 1995)

If so, certainly accreditation of certification organizations will continue to play an important role in the harmonization and application of standards. Several key questions, however, remain. To what extent will citizens, consumers, and nongovernmental organizations continue to influence the development of standards and the certification and accreditation processes? Who will be the ultimate victor in the "struggle for standards"? Will accreditation be able to effectively integrate global principles with local decision-making processes? And finally, and perhaps most vexing, to what extent will the proliferation of global principles, forest management standards, certification organizations, and accreditation programs actually contribute to the maintenance or even improvement in the health and status of the world's forests?

References

Barrett, Richard. 1993 *Quality Manager's Complete Guide to ISO 9000.* Englewood Cliffs, NJ: Prentice-Hall.

Cabarle, Bruce, et al. 1995. "Certification Accreditation: The Need for Credible Claims." *Journal of Forestry.* Vol. 93, no. 4 (April 1995).

CPPA. 1995. "Sustainable Forestry: Toward International Certification, A Canadian Approach." Brochure printed by the Canadian Pulp and Paper Association. Rexdale, Ontario.

Droogsma, W. D., et al. 1991. "Legal Means for Restricting the Impact of Nonsustainably Produced Timber: Aspects of International and European Law." Center for Environmental Law. University of Amsterdam.

FSC. 1994a. *Principles and Criteria of Natural Forest Management.* Mexico. June 1994.

FSC. 1994b. "'Guidelines for Certifiers." In *FSC Statutes.* Forest Stewardship Council, Avenida Hidalgo 502, Oaxaca 68000, Oaxaca, Mexico.

FSC. 1995. "Manual for Evaluation and Accreditation of Certification Bodies." Draft 3.0. Oaxaca, Mexico.

Hutchins, Greg. 1993. *ISO 9000: A Comprehensive Guide to Registration, Audit Guidelines and Successful Certification.* Oliver Wight Companies, Essex Junction, Vermont.

ISO. 1994. Draft ISO Guide—General requirements for assessment and accreditation of certification/registration bodies. Geneva, Switzerland. ISO/CASCO 226 (Rev. 2). November 1994.

Pinkham, Fred O. 1952. "The National Commission on Accrediting Progress Report." In *Proceedings of the Northwest Association of Secondary and Higher Schools.* Seattle: Northwest Association of Schools and Colleges.

Read, Michael. 1991. "An Assessment of Claims of 'Sustainability' Applied to Tropical Wood Products and Timber Retailed in the UK." Report by the Flora and Fauna Preservation Society, United Kingdom.

Wilson, Woodrow. 1907. "Schools and Colleges." In *Proceedings of the Middle States Association of Colleges and Secondary Schools*. Philadelphia: Middle States Association of Colleges and Schools.

Young, Kenneth E., et al. 1983. *Understanding Accreditation*. San Francisco: Jossey-Bass Publishers.

Chapter 5

Forest Assessment

Kate Heaton and Richard Z. Donovan

Techniques of forest assessment for certification are best appreciated after considering the overall forest management context. This chapter, therefore, describes current forestry practices and patterns of logging, discusses a range of desirable and undesirable forest management practices, presents forest-level certification assessment procedures, and concludes with perspectives on the benefits and difficulties of forest assessment processes.

Certification is a rapidly evolving field; forest assessment techniques represent one area where there is great potential for innovation, with opportunities for providing greater assessment accuracy, reduced subjectivity, and increased efficiency. Current assessment practices include a mix of good science, systematic fieldwork, and knowledgeable judgment. Though virtually everyone is striving for scientific knowledge to provide the determining framework for guiding certification assessments and decisions, it is not realistic or practical to expect this in the near future, since "best available technologies" (or "best management practices") are often still based on insufficient scientific foundations.

Logging, Clearcuts,
and Sustainable Forest Management

For better or worse, many public discussions of forest management start with logging and clearcuts. Logging and clearcuts are clouded by negative environmental perceptions, and often rightly so; however, it would be incorrect to imply that all logging is highly destructive. In some cases, timber management may be the most environmentally desirable land use, especially when viewed as an economic alternative to more destructive practices such as conversion of land to agriculture and pasture. When evaluating the appropriateness of logging techniques (including clearings) certifiers must consider the objectives and justifications for using them, the potential impacts, and in many regions the relative merits of competing land uses.

Given the negative press focused on clearcuts, there is a common misperception that if an operation is not clearcutting, it is good and, therefore, must be certifiable. In fact, concerns about clearcuts have reached such a level that many operations apply for certification based on the mere fact that they remove only a selected number of trees. While such "selective harvesting" may be viewed as generally positive from some environmentalists' perspectives because of its relatively low impact, it does not alone constitute good and certifiable forest management. Conversely, while uncontrolled clearcuts must be discouraged, acceptable forest management must provide for the controlled use of openings that are appropriate to the local ecology and sufficient to regenerate forest stands, because some species and forest types will not reproduce after selective harvests.

Attempting to Define
Sustainable Forest Management

There is no single definition of sustainable forest management; it can be defined in terms of timber sales or cash flow by the forest owner, minimal disturbance by the environmentalist, and stable socioeconomic conditions by the social scientist. Scientific data do not yet support a single consensus on definition of biological sustainability, especially given regional variations in ecology; the same is true for socioeconomic sustainability. Nonetheless, certification can differentiate many of the key elements of good and bad forest management, and can begin to give clarity to the path toward sustainability.

Defining what constitutes certifiable forest management is complicated by the fact that there is also a wide gap between ideal forest management as might be described from an academic or environmentalist perspective and that which is actually occurring on the ground. This requires, therefore, that certification be viewed as a process of continual improvement beyond initial minimum thresholds. Initial certification thresholds must be reached, and procedures put in place for continual improvement.

Certification standards include a mix of both process criteria (e.g., monitoring systems are in place) and prescription criteria (e.g., no harvesting on steep slopes of defined gradients). They address impacts at the site, stand, and landscape levels. Certification standards must include sufficiently prescriptive criteria to protect existing forests while allowing managers and certifiers enough flexibility to pursue appropriate management decisions on a case-by-case basis. In identifying acceptable levels of ecosystem disturbance, certifiers must consider the temporal and spatial scale of proposed changes (Is the proposed change occurring over one or one hundred years, or on 10 percent or 100 percent of the land base?), their intensity (Is harvesting a clearing or selection cut?), and their relationship to competing land uses and the maintenance of ecosystem function.

Reaching a certification decision requires analysis of the ecological, socioeconomic, and silvicultural implications of any management system that is being implemented. Usually, there are trade-offs between silvicultural, ecological, and socioeconomic priorities; but in general certifiable management must balance biological conservation, social, and timber production goals in, for example, an integrated landscape management approach.

Minimum requirements for forest management certification, by programs such as Smart Wood or the Woodmark Program of the Soil Association, are:

1. Long-term security for the forest resource
The forest must be committed to long-term forest management, or there must be strong indications that it is not vulnerable to permanent clearing or change in land use. For example, if a natural forest is going to be cleared in three years for citrus plantations, it will not qualify for forest management certification, even if the logger is doing reduced impact logging. In the case of tree plantations, with few exceptions, it is unlikely that an operation will be certifiable if large expanses of natural forest were recently cleared by the current forest managers in order to establish the plantation.

2. Overall compatibility with general or region-specific certification standards
Certification is based on written standards or guidelines. Separate guidelines often exist for natural forest management and plantations. Each certifier usually has its own guidelines. According to the Forest Stewardship Council's Principles and Criteria, acceptable certification guidelines must address: (1) environmental impact (e.g., biodiversity, soil erosion, watershed protection); (2) sustained yield impact (e.g., silviculture, growth and yield, replanting, regeneration potential); and (3) socioeconomic impact (e.g., respect for local people and workers, community stability).

A challenge for the field of certification assessment is to begin to define the key elements of sustainability regionally through regional standards development. (See Chapter 5.)

3. Existence and implementation of a good forest management plan
In virtually every case, no operation can be certified unless it has a good forest management plan that is both written and implemented. The level of detail required in the plan is related to the size and sophistication of the operation. Typical parts of a certifiable forest management plan include:

- An inventory of the forest resources
- Definition of cutting cycles or rotations lengths
- Mapped annual cutting areas and management units
- Annual allowable cut defined so that growth balances offtake
- Appropriate ecologically based silvicultural management techniques prescribed (e.g., protection of seed trees, replanting, vine cutting)
- Appropriate environmental management techniques prescribed (e.g., directional felling, protection of important wildlife trees)
- A plan for monitoring forest growth and regeneration
- Identification of conservation zones for protection of fragile areas (e.g. waterways, slopes, or biologically diverse areas)
- Strategies for minimizing or eliminating use of chemicals
- Roads and skid trails planned to reduce residual impacts

Certifiable Natural Forest Management

There are two general types of natural forest management: (1) selective harvesting of a few timber species followed by seedling regeneration into the existing stand, resulting in uneven-aged forests; and (2) intensive cuts (including clearcuts), usually followed by flushes of seedling regeneration that typically results in the development of relatively even-aged stands. The number of trees removed during logging operations varies by forest type, region, and available markets.

Logging is typically selective in forests with high species diversity, where only a few species are considered to be valuable timber. This is the case in the many mixed hardwood forests common in the United States (as in the midwest and northeast) and tropical Latin America (as in most of Central and South America). Selective harvesting of commercially desirable species results in varying degrees of damage, ranging from removal of only two or three trees per hectare to more extreme damages when one-third to two-thirds of the remaining trees are left damaged, canopy cover is reduced by 50 percent, and roughly 10 percent of the forest floor is left compacted by logging roads and skid trails (Uhl and Guimaraes 1989). In many Latin American tropical hardwood forests, a common scenario might involve harvesting 2 to 10 percent of the approximately 100 to 150 potentially harvestable stems per hectare.

Clearcutting is usually employed where high-value species (such as Douglas fir, *Pseudotsuga menziesii,* or loblolly pine, *Pinus taeda*) occur in dense, uniform, and typically even-aged stands, as in the Southeastern or Pacific Northwest portions of the United States. Similarly, in Southeast Asian tropical moist forests the cut is typically very heavy where stands contain a high percentage of large and highly valuable Dipterocarp species. Unless very well planned, such heavy cuts can resemble clearcuts.

Certifiable natural forest management must allow for openings sufficient to promote regeneration of species appropriate to the ecosystem, because there are instances where low-intensity harvesting with minimal disturbance can actually contribute to the demise of certain species from the forest. For example, species such as bigleaf mahogany (*Swietenia macrophylla*) and eastern white pine (*Pinus strobus*) are "light loving" and their regeneration depends on opening the canopy enough to allow in light. Mahogany regenerates well in wind-tossed hurricane zones in Central America. White pine regenerates well after sweeping forest fires in the northern United States. Over time, such light-loving species will tend to disappear from forests if only selective harvesting takes place, since such harvesting will favor the regeneration of other species whose seedlings are capable of growing in the shade under the forest canopy.

Consequently, some forest management intervention may be necessary to maintain these species (or forest types), especially since some widespread disturbances, such as forest fires, have been actively suppressed by humans. In many areas, potential environmental problems associated with large-scale, intense harvests and clearings are well known. Such problems may include loss of wildlife habitat, soil erosion, and degraded watersheds, or serious changes in the quality of forest regeneration. Certification requires that such destructive practices be controlled, but this must be done carefully. Attempts to set blanket restrictions on the intensity of harvests may be arbitrary, with potentially negative implications for some species or forest types.

One of the main problems in forestry today is the dominance of completely unmanaged logging. In the United States alone, the Society of American Foresters estimates that over 60 percent of all timber harvests are not supervised by foresters. Unmanaged logging can be characterized as follows:

1. No forest inventory

2. No forest management plan

3. High-grading (selective over-harvesting) of preferred species and geneti-
    cally superior trees

4. Multiple unplanned reentries into previously logged areas, interfering
    with regeneration

5. No designation of conservation zones such as steep slopes, streams, or
    biologically important areas

6. Uncertainty over whether the area will remain in forest or be converted
    to other uses in the near future

Poor management can often be found even where low impact harvesting
systems are employed, but the long-term viability of the timber resource is
compromised nonetheless due to lack of management. For example, in unmanaged mixed hardwood forests, such as those commonly found in the
Latin American tropics, high-grading of the most commercially desirable
species is a common practice. Loggers rove the land hunting for the best
formed specimens of the most valuable trees, such as bigleaf mahogany (*Swietenia macrophylla*) or tropical cedar (*Cedrela odorata*). Since these high-value
trees usually represent only a small fraction of the available stems (perhaps 2
percent), they may be quickly depleted from an area. Such high-grading of
the healthiest, best formed, and favored species of trees, whether few or
many, may lead to genetic erosion of the forest and a long-term decline in the
quality of the resource. The less desirable species and misformed, genetically
inferior, diseased and dying trees are left to reproduce and take up valuable
growing space. The forest may then be considered to be diminished in value.
In the worst case, land managers may then seek to convert it to other more
short-term, financially lucrative land uses such as agriculture and cattle pasture. In such situations, certification may assist in discouraging uncontrolled
logging by introducing the concept of long-term "resource management" into
the practices. (See Box 5.1.)

In summary, for sustainable natural forest management initiatives to succeed in the long run, managers must inventory the forest, define conservation
zones, divide the productive land base into annual cutting blocks (e.g., 25
units) that supply wood over a full cutting cycle or rotation (e.g., 25 years),
and define the annual allowable cut for each block. Loggers must take only
what the forest has to offer in each annual cutting block, as defined by the
natural species composition, rather than high-grading preferred stems. A
controlled number of trees is designated for harvest, so that offtake is balanced with growth over the cutting cycle for any given area. The annual allowable cut is defined by the forest's overall growth rate and individual tree
species' growth rate and abundance; in turn, these are determined by natural
ecological processes and human actions through harvesting or other activity.

Box 5.1

Reduced impact logging as a starting point for certification

Starting in 1992, Innoprise Sbn. of Sabah, Malaysia, with financial support through New England Power of the United States, conducted a three-year pilot project to test the impact of reduced impact logging techniques in tropical moist forest (Putz and Pinard 1993). Based on research conducted on site, techniques of reduced impact logging can decrease damages to the forest (residual stand damage) by 50 percent, thus helping to protect an increasingly scarce and valuable resource. The key techniques included were careful skid trail and forest road planning, preharvest severing of vines (climber cutting) that join unwanted trees to those selected for cutting, directional tree felling, and very controlled use of tree harvesting equipment or "skidders" and bulldozers (e.g., keeping equipment on skid trails and out of forest, aggressive use of cable to pull logs to skid trail, reduced blading of soil). Similar positive experiences with reduced-impact logging techniques have been demonstrated in research projects in Surinam, Brazil, and Costa Rica, and in commercial operations in Australia. Such techniques are now prerequisite for certification in most natural forest management operations.

In most cases, establishment of a viable annual allowable cut will require systematic forest monitoring at the levels of individual tree species and forest type.

In most operations, policies need to be developed and implemented to prevent the erosion of the quality of the timber resource. It is very important to define appropriate criteria for tree selection. Operations should consider the feasibility of employing "risk and vigor" tree selection criteria, instead of relying solely on the minimum diameter limits that are commonly used. "Risk and vigor" selection removes trees that are at risk of dying before the next cutting cycle and leaves those that are vigorously growing and adding volume whose value may be best captured in the next cutting cycle. One of the outstanding examples of risk and vigor selection is the Menominee Tribal Enterprise's forest in Wisconsin. As a result of this practice, the forest is very healthy and diverse, with trees of all ages and sizes, including 300-year-old white pines (*Pinus strobus*), and a standing merchantable timber volume that is significantly higher today than it was in the past.

While it may be important to remove some of the over-mature and hollow trees to create growing space, it is also essential to recognize the importance of maintaining an ample supply of old, dead, and downed trees as "dens and snags" for wildlife habitat. One estimate for temperate forests suggests leaving at least two to four den and snag trees per acre (Hunter 1990). Dead and downed trees are also very important as sources of nutrients and organic matter, which positively affect forest growth and soil stability.

Certifiable Tree Plantations

In the case of plantations, certifiable attributes include many of the above characteristics, as well as the following:

- Operations are designed with a combination of timber production and biological conservation goals.
- Degraded lands, such as pasture, are reclaimed via replanting.
- No new primary or advanced secondary natural forests are cleared for plantation establishment (though previously established plantations may be acceptable under certain circumstances).
- An effort is made to plant mixed native species and suitably adapted exotic species.
- Remaining forest fragments on the property are protected for conservation and joined with planted corridors of mixed native species as appropriate.
- Plantations are used as a first step toward restoration of lands to natural forests for timber management and other objectives (usually desired, not required).

An excellent example of a certifiable plantation is seen in a certified operation called Tropical American Tree Farms in Costa Rica. Here the owners have skillfully balanced conservation with production goals, using an "integrated landscape management" approach. They have reclaimed degraded pasture by planting 60 percent native species and 40 percent teak trees as a crop; permanently restored denuded steep hillsides in native cover for soil and watershed protection; committed to preserving the remaining forest fragments on the farm; and planted native species corridors to join the forest fragments.

Forest-Level Certification Procedures

Having explained the basic goals of good forest management and certification, it is important to consider the processes and procedures used for forest management assessment. Major steps in the source certification process include:

1. Discussion between applicant and certifier
2. Application submission with supporting documentation

3. Assessment budget agreement

4. Evaluation team assembly and assessment planning

5. Field assessment by an interdisciplinary team

6. Assessment report write-up

7. Confidential peer reviews of assessment by two or three independent specialists.

8. Terms and conditions for certification defined by certifier and agreed to by applicant

9. Certification decision by certifier

10. Processing of final payments, certification contracts, certificates, and press releases

11. Annual and random audits

Certification is not a rapid process. It requires a lot of careful work and coordination by many people. Normally, certifications take four to six months to complete; operations that need to make improvements prior to certification will take longer. Medium-to-large sources can require 40 to 60 person-days of labor to certify, and work is punctuated by waiting time (for document exchanges, scheduling, processing, and so on).

Forestry Operation Assessment Methodology

Source assessments are typically done by three-person interdisciplinary teams whose expertise mirrors the composition of the certification guidelines (e.g., a forester, an ecologist, and a social scientist). During field evaluations, it is recommended that certifiers use local talent wherever possible as a complement to staff evaluators. Local evaluators are desired in order to gain local perspective, build in-country certification capacity, facilitate follow-up and monitoring, and reduce costs.

Source assessments typically last one to three weeks and include (1) field site and facilities visits; (2) meetings with the operation's management; (3) meetings with affected parties, such as government forestry officials, company employees, local communities, environmental groups, and academics; (4) consideration of local and national forestry laws affecting the operation; and (5) tabling and discussion of the guidelines for adaptation to local conditions as needed.

Field assessments cannot and need not occur on every hectare in a concession or management area. Assessments are focused most intensely on a sample of the current and past harvesting blocks. Following are some of the commonly employed field assessment techniques.

Assessors visit a "chronosequence" of sites representing a sample of the various stages of logging and regeneration, such as currently harvested blocks, blocks harvested two or three years ago, blocks harvested five years ago, and blocks harvested ten or more years ago. Emphasis is often placed on assessing the current harvesting blocks.

In the case of larger commercial operations, an airplane overflight is recommended for the entire concession to provide a broader perspective on the area and management. This is useful for verifying the overall impact of the logging and road network and for determining whether any unauthorized activities and encroachment are occurring. An overflight is also one of the best ways to see areas that have been taken out of production (such as wetlands, steep areas, and conservation zones) and verify the quality and stocking of the unlogged production forests that are scheduled for future harvesting.

Transects or other sampling systems may be employed to determine the incidence of occurrence of desirable practices (such as vine cutting and directional felling) or undesirable practices (such as residual stand damage). To organize this, all properties where management has taken place should be stratified according to the supervising forester, harvesting contractor or team, equipment used, soil type, forest type, time of year when the harvest was conducted, or other similar determining factors. When this stratification is combined with the chronosequence described above, a strategy can be devised to visit a representative sample of management sites throughout the operation.

One of the most important aspects of the assessment is determining that the average biomass offtake is equal to or exceeded by forest biomass growth during the defined cutting cycle. Drawing on scientific growth studies in Surinam, South America, the rate of commercial timber species growth for Latin American tropical moist forests is often estimated to be about one cubic meter per hectare (m^3/ha) per year (Jonkers 1987). Thus, in an area visited on a cutting cycle of once every 15 years, an operation can harvest up to 15 m^3/ha and still be in balance with expected biomass growth and replacement during the cutting cycle. The m^3/ha per year is a gross estimate; in the long run, it should be verified through the collection of data on actual growth rates from permanent plots to be set up within the management area.

Estimating allowable offtake becomes much more difficult for individual species. There is often little information (or time) to verify the individual growth rates of species and compare them with offtake rates. Therefore, it is important to make sure that the cut is being distributed across many species and age classes, to avoid undue impact on any species or generation of timber. Operations that rely on too few species will be asked to increase their utilization of lesser known species, while reducing selection of favored species. As an additional safeguard, inventory data should be assessed to assure that very rare or vulnerable trees are not being over-harvested. Species with very low frequencies of occurrence in the forest (e.g., fewer than one mature tree per 25, 50, or 100 hectares) or that appear on the local or national threatened and endangered species lists should be excluded from harvesting.

In active forest management zones, several measuring instruments are commonly used. Measuring tapes are used to gain perspective on the extent of forest loss due to roads, logging decks, and skid trails, and to determine proximity of harvesting activities to stream or river buffer zones; for example, stream buffers should be approximately equal to the width of the waterbed on each side. Clinometers are used to determine if logging is occurring on slopes that are too steep according to local laws and the certifier's criteria (e.g., greater than 35 degrees) or if road grades are too steep (e.g., greater than 25 degrees).

The certification evaluation team is always expected to follow the guidelines and the assessment procedures in order to assure that there is consistency in the factors considered for each operation, but there is also a firm respect for and clear recognition that team members should use their professional judgment in ensuring that an appropriate range of factors is considered in evaluating an operation.

Certification cannot be expected to assess all areas or solve all problems. There are limits to what it can assess and achieve due to cost, time, and information constraints. It is common in the tropics to carry out forest management without sufficient information about wildlife and endangered species in a harvested region, and without knowing the impacts of logging on the flora and fauna. This also occurs in temperate zones. Information on the growth rates and regeneration habits for many inventoried tree species may be lacking. This may make it difficult to decisively protect individual species from over-exploitation, except as indicated by the practical measures described earlier.

There is continuous debate about how involved certifiers should get in social issues. Some see certification playing a greater role in environmental issues than in social ones. Experience in the marketplace indicates that consumers are concerned about social issues, such as the rights of indigenous peoples or how a company treats its employees. In most cases it appears the public understands that assessment of social conditions must first be considered within the regional and cultural context, not the international context. Current FSC policies require that certification should assure that there are no obvious abuses and that basic human rights are respected (indigenous land claims, worker safety, and payment at or above the regional minimum wages). Social benefits beyond basic rights (employee profit-sharing plans, full health coverage, and so on) may be recognized, encouraged, and favorably reflected in certification scoring, but it may unfortunately be beyond the scope of certification to routinely or definitively establish these benefits as requirements at this time. (See Box 5.2 for more information on certification and community forestry.)

Box 5.2
Certification and community forestry

Proponents suggest that certifiable management will result in more stable, long-term sustainable development through forest management systems that stabilize the flow of timber, improve the future timber crop, provide long-term jobs, foster sustainable community development, and create value that serves as an incentive to maintain property in forest cover. Where land is under population pressure to yield an economic return, it may even save forests. The potential for forest management programs to save forests is powerfully illustrated by the words of one Honduran "campesino" (peasant) interviewed by Smart Wood in 1993:

> I used to be a slash and burn cultivator. Now I feel like a criminal when I think of all the forests that I have cleared. For every tree that I now harvest for timber, I get the same value as I would get for clearing an entire hectare of forest to plant corn and beans as a cash crop. I make three times the agricultural wage of $2.50/day by harvesting trees, and this improves the standard of living for my family. Were it not for harvesting trees, the 3,000-hectare forest area where my community has lived for 13 years would have been cleared for agriculture by now.

The successful implementation of community forest management systems in such areas is complicated by many factors. For example, many of the species in the tropical forest where this farmer lives are lesser known species for which markets are poorly developed. Marketing problems, therefore, create a strain on the implementation of management systems because if people cannot sell the species they harvest, they may have no incentive to continue following the management plan, or to maintain the forest. Thus, for true sustainability, it will be crucial to develop markets for the lesser known and utilized species. Fortunately, certification is beginning to draw attention to species that might otherwise be ignored by the market, even though in many cases their beauty and workability may be as good as that of favored woods.

Certification Decision Making—Scoring and Categories

There are various approaches for reaching final certification decisions. For example, to be considered for certification in the Rainforest Alliance's Smart Wood Program, operations must receive an average score of at least 3 (neutral) on a scale of 1 to 5 in each of nine different certification subjects (land tenure, environmental impact, community relations, and so on). In the Scientific Certification Systems Forest Conservation Program (SCS), operations must receive a score of at least 80 percent out of 100 percent in each of three general categories: sustainable harvest, ecosystem health, and community benefits and financial considerations.

If a forest management operation attains passing marks in the certification process, it is assigned to a specific certification category depending on its score. In practice, there are currently three basic source certification categories, although the exact names may vary according to certifier: (1) sustainable—or state-of-the-art well managed; (2) well managed; and (3) precertified, or transitional.

To date, with the exception of Collins Pine in California, which was certified "state of the art and well-managed" by SCS, certifiers have certified operations as "well managed" due to a lack of data to substantiate long-term claims of sustainability. Mutually agreed-upon conditions for improvement are virtually always attached to the certifications. Because of the wide gap that exists between ideal forestry management and actual practices on the ground, certifiers understand that certification must foster a process of realistic improvements in operations over time, with progress being monitored during annual (in some cases more often) audits.

Scoping assessments or precertification assessments are often done for sources that are not ready for certification, but wish to learn what is needed to comply. Once the applicant has satisfied a mutually agreed-upon timetable of improvements, certifiers will reassess the operation for certification. Precertification cannot be used for product promotion or advertising, but may be used to support the forestry project's fund-raising efforts. This is significant as more and more banks and investors consider the environmental impacts and sustainability of projects when deciding whether to grant loans and investments.

Conclusions

Certification can be a very powerful tool for measuring progress toward achieving the very important and elusive goal of sustainable forest management. Certification assessment techniques capitalize on experience gained throughout the forestry sector, such as watershed evaluations and rapid rural appraisal. As experience and research with assessment techniques grow, so will the effectiveness and efficiency of certification procedures.

Certifiers today are learning that there is a very real gap between actual and ideal practices. Therefore, not only must certification be used as a tool to foster a process of realistic improvements in operations over time, but it must also be used as a means to help society define what it can reasonably expect of operations and from certification. In sum, as a young art and science, certification must remain an iterative process of defining what can and should constitute good forest management.

References

Hunter, Malcolm. 1990. *Wildlife, Forests and Forestry* (ch. 10). Englewood Cliffs, NJ: Prentice-Hall.

Jonkers, W. B. J. 1987. *Vegetation Structure, Logging Damage, and Silviculture in a Tropical Rainforest in Surinam.* Agricultural University, Wageningen, The Netherlands.

Putz, Francis E., and Michelle A. Pinard. 1993. "Reduced-Impact Logging as a Carbon-Offset Method." *Conservation Biology* vol. 7, no. 4, December 1993.

Uhl, Christopher, and Ima Celia Guimaraes Vieira. 1989. "Ecological Impacts of Selective Logging in the Brazilian Amazon: A Case Study from the Paragominas Region of the State of Para." *Biotropica* 21(2): 98–106.

Chapter 6

Chain of Custody

Michael Groves, Frank Miller, and Richard Z. Donovan

Chain-of-custody auditing is the second major component of forest management certification, complementing a forest assessment (see Chapter 5). Chain of custody can be defined as "an unbroken trail of accountability that ensures the physical security of samples, data, and records." Chain of custody is a critical element of any certification system since it provides the ultimate link between the "consumer" and the "producer." The term consumer is used here in a broad sense to indicate individuals, retail companies, or suppliers in countries where wood and wood products are distributed, sold, or used. The producer may be an individual landowner; a community organization (such as a cooperative); a company that manages forests, brokers products, and/or processes wood; or a government.

The desire by consumers to assure themselves that the wood or other forest products they sell or purchase are from well-managed sources is one of the key motivations underlying attempts to develop credible certification systems. In this context, if the consumer seeks assurance, the responsible producer provides this assurance through a certificate stating that the product does indeed derive from a certified, well-managed forest. In certification, the burden of proof is on the producer to develop a reliable or confident system for ensuring a clean and clear chain of custody. In practice, such systems are often developed, or at a minimum fine tuned, after interaction with a qualified certifier.

What Should a Chain of Custody Look Like?

Ideally, a secure chain-of-custody system will have the following basic elements: (1) physical evidence (such as documents, tags, and labels) that the goods originate from a particular source; and (2) an "auditable" data recording and communication system that runs in parallel with, and links to, the physical evidence identifying each product.

The chain of custody itself will consist of a number of links—the number depending on the range of sources, the complexity of the manufacturing process, and the type of market into which the product is sold.

A somewhat detailed model of the wood product chain showing links and a mode of transfer across the links, from the standing tree to final use, is shown below. Note that this model is general and is not intended to cover all possible material and process flows.

Critical Link	Mode
Standing tree to a log at stump	chainsaw, harvester, axe
Log transported from stump to roadside	skidder, tractor, horse, helicopter
Log at roadside to the log pond	truck, rail, water
Log at log pond transported to industry log yard	truck, barge
Processed goods transported to point of export	truck, ship
Import point where bulk product is broken down	truck
Transfer of goods to final point of sale	truck
Final point of sale to the end user	

Other links may be added—for example, when manufacturing or processing is on a multisite basis. It is quite typical for one mill to produce dimensional lumber and another mill to further process or finish products. Alternatively, links may be removed—for example, when goods are distributed locally.

If credibility is to be maintained, the chain of custody must remain intact and clear throughout, particularly as physical responsibility for the goods change (e.g., from sawmill to transporter to wholesale or retail outlet). Essentially, chain of custody is a stock control exercise, which requires the goods to be secure and requires transparency for ease of inspection. The latter is critical, since inspection is the vehicle through which the consumer is assured that the goods derive from a well-managed or indeed any specified source. Such inspections can take place on a programmed or random basis.

Chain-of-Custody Technologies and Methods

The key to successful implementation of a chain-of-custody audit, as it is for forest assessment, is the application of appropriate technologies and methods for all stages. Technologies must be dependable, have an acceptable cost per output, and be easy to use. Methodologies should be clear and auditable.

Matching the chain-of-custody methodology to scale (business size) and scope (geographic reach from producer to market) of the operation are critical. For example, a publicly listed company with many timber and manufacturing interests, selling into a variety of national and export markets, will require a different approach than a community-based forest management operation supplying through one intermediary to a regional or national (nonexport) market. The multisite, multimarket operator will require a certification program with a global scope. If goods are being shipped from multiple sites to multiple markets, sending inspectors halfway around the world to audit the chain of custody is usually not a cost effective approach. Systems must be set up that use locally based staff to conduct auditing.

Obviously, it is essential that the certifying organizations provide a cost-effective chain-of-custody audit service to any client that it chooses to serve. Two different models currently exist for accomplishing this. First, there are international for-profit firms, such as SGS-Forestry or Scientific Certification Systems, which have either full offices or designated representatives in different regions and countries. Second, there are networks of affiliated organizations, such as the Smart Wood Network. In the latter case, contact persons who serve as regional representatives or affiliates for the Smart Wood Program have been established at collaborating nonprofit organizations in many countries.

An initial consideration must be the selection of suitable identification techniques for verifying the source of a material coming out of the forest. Traditional techniques include paint, hammermarks, labels, and latschbacker tags. Paint is the most common marking technique and includes paint daubs possibly indicating forest of origin, species, and volume. Hammermarks are commonly used to mark logs to verify measurements taken at the time of felling. Labels are similar to paint because they can display the same information; however, unlike paint they can be repositioned above the water surface if the logs are rafted. Latschbacker tags are numbered plastic labels that are applied to the log with a single blow from a specially adapted hammer. They can carry a single number only; information pertaining to the log must be recorded elsewhere.

Some new techniques are also being used by some industries, or being tested in various situations, including barcoding, radio-frequency identification devices, and touch memory. Barcoding is rapidly becoming a standard technique in many areas of business. Barcode labels are designed to be machine readable via a barcode scanner attached to a personal computer. When applied to log tracking, they provide a rapid and accurate technique for obtaining data on the logs. They also have the advantage of being extremely secure, with each barcode providing a code unique to each log. Barcode labels have been used operationally for log identification in Australia and New Zealand, in addition to pilot efforts in Indonesia. (See Box 6.1.) The labels are easily manufactured and can be of virtually any size.

Radio-frequency identification devices are essentially preprogrammed computer chips that incorporate a small area through which data can be received or transmitted. The whole device fits into a glass cylinder about 15 millimeters long and four millimeters in diameter. The device can be inserted into the end of a log and read using a hand-held radio-frequency identification device reader. Although tested, these devices are generally too expensive to be widely used for log tracking.

Touch memory consists of a battery-powered computer chip in a circular metal housing with a diameter of 20 millimeters. It has the advantage of being able to store the equivalent of one page of text and can be read easily via a touch memory reader attached to a personal computer. (See Box 6.2.) The touch memory unit is robust and can be reused; one unit can be expected to last for three years.

Box 6.1
Chain of custody for export control in Cameroon

In Cameroon, SGS-Forestry has installed, at the behest of the government, a system of labeling, scaling, and grading for 1.5 million cubic meters of annual log exports. This system is based on the use of barcodes and hand-held barcode readers, which allows the inspection of approximately 7,000 cubic meters per day. The code is married to information on species, exporter, forest of origin, log dimensions, and quality. This information is downloaded into a central database where inspection results are reconciled against shipping specifications supplied by the exporter and all export details are checked. The analyzed and reconciled information is subsequently provided to the government. As well as producing the associated software, training, operating manuals, and market information, SGS-Forestry has been asked to expand the scope of the project and develop the system for tracking logs back to their sources. The eventual aim is to undertake assessments of forest management quality at these sources.

Box 6.2

Chain-of-custody research in Indonesia

In September 1993, SGS-Forestry in Indonesia was funded jointly by the Indonesian Ministry of Forestry and the European Commission to test a variety of techniques for tracking logs from the stump to the mill. During the Indonesian project, touch memory units were used to label and track finished products from Kalimantan to the United Kingdom. The model for a verifiable chain of custody from the stump to mill used barcodes, handheld computers, and satellite modems to capture and transmit data. The trial in Indonesia tested the techniques described from the standing tree to the roadside or log yard, to the riverbank or log pond, through the sawmill, to the point of export, to the point of import and the receiving warehouse.

The second component of the chain of custody is the means (or technique) to capture, communicate, and audit data associated with the goods being processed and transported. Recording information on paper is of course the traditional and still principal method in most situations, despite the increasing use of the newer technologies such as barcoding.

Written lists and forms can, however, become unwieldy where the number of units is high—for example, where thousands of logs are being loaded onto a ship or hundreds of bundles of processed goods are loaded into containers. Thus computerization is rapidly being explored as either a stand-alone or complementary approach. It has the advantage of being able to store large volumes of information that can be readily accessed and cross-referenced. The data may also be transmitted instantaneously via telephone lines, fiber optic cable, or satellite, or be printed.

Whatever technology is chosen, a credible system will also establish a series of key inspection points along the chain that will provide the necessary verification that the system is working.

Key Stages in Chain of Custody

Though there are many different potential scenarios for a full chain of custody, the flow of product from the forest to the consumer can be divided into the following three key stages: (1) from the forest to processing/manufacturing mill; (2) during the processing/manufacturing stage; and (3) from processing to the market.

Stage 1—From Forest to Mill

The first stage of chain-of-custody auditing usually involves developing a credible system for tracking product from the forest to a processing center. Such a system must meet the following requirements:

1. Provide up-to-date information on volumes of standing timber targeted for logging, volumes felled, volumes stored at landings and log yard, and volumes delivered to the mill.

2. Allow logs to be traced back to their origin, thereby determining their source.

3. Provide accurate and timely information on log volumes and species composition at the felling sites as a basis for calculating fees and royalties.

4. Allow the identification of log origin (to block and plot level) in order to identify logs felled illegally.

Stage 2—During Processing or Manufacturing

The second stage of chain of custody is during processing and manufacture. The following sequence traces the process flow from the forest to the point of export for a typical wood product. In this case, an emphasis is placed on monitoring material flowing through the mill and subsequent molding and assembly operations. The chain-of-custody audit sequence for a chain-of-custody auditor might be as follows:

Main offices. Review letters of credit, bills of lading, and container loading reports; auditor also quantifies the amount of product shipped, reviews work orders issued for each product, and obtains a shipping schedule.

Accounts department. Reviews documentary credit requirements, invoices, receipts issued by the manufacturer, payment vouchers issued by the manufacturer, and information on the supplier stored by the manufacturer.

Manufacturer's kiln-drying department. Reviews delivery orders from the supplier, information on each bundle, kiln-drying schedule for each bundle, rebundling records, and internal bills issued to the store and molding departments.

Supplier's office. Reviews contract to supply the manufacturer, delivery orders, invoice to the manufacturer, and receipts issued by the manufacturer.

Forest management office. Reviews contracts between forest manager and supplier to manufacturer.

Manufacturer's wood store. Reviews internal bills issued by the kiln-drying department, verifies bundle details and delivery records.

Manufacturer's molding department. Reviews records of input from the wood store, undertakes a random check of bundles from the store, reviews daily input records and daily product assembly records.

Final analysis. Adds up all the figures and compares input data with output data, taking into account appropriate wood conversion factors for different products (how much wood goes into a chair, a table, or a molding product).

The types of documentation and procedures that are audited in this example should be compatible with the requirements of a quality management system, such as the widespread International Organization for Standardization (ISO) 9000 quality management certification. Implementation of quality management systems can have direct benefits to the forest and wood product industry. ISO 9000 and the emerging environmental management system standard, ISO 14000, provide a complementary framework for managing operational issues that might affect organizations at all points in the wood value chain, from the producer to the consumer.

A number of companies are already using fairly sophisticated barcoding systems for inventory control. For example, Herman Miller, Inc., an office furniture manufacturer with headquarters in Michigan, receives veneer originating in Africa from suppliers that have a bar code attached to each individual sheet of veneer, identifying species, source location, and other shipping-related information. Colonial Craft, a molding and millwork company also in the United States, uses a series of multicolored tags that are also barcoded.

Stage 3—From Final Processing to the Marketplace

Once processing has been completed, the product may be packaged, stamped with a label or identifying tag, and/or shipped in a banded bundle. With few exceptions, every product shipment is accompanied by an invoice or bill of lading. If the product is moving into an international market, each product load may be accompanied by a GATT Form A Certificate of Origin, which lists the quantity and stock numbers for all items, commercial value of the product being shipped, point of origin, and final destination. All of these documents provide an auditable trail, which can then be checked at each critical link of processing or transport.

Different companies and certification organizations have unique labels or other ways of presenting the certified product to the public. These also provide a method for chain-of-custody tracking. For example, Seven Islands Land Company has worked with a flooring company to produce flooring boards that have "SCS certified" etched into the bottom of the floorboard. Collins Pine sells SCS-certified "Collinswood" in the form of shelving that is in shrink-wrapped packages with a descriptive brochure included. Smith & Hawken sells Smart Wood–certified furniture products with hangtags that include the Smart Wood label and a limited amount of background information. The more complete the packaging, the more conclusive the final identification of a certified product can be.

Working Backwards—
Designing Chain of Custody from the Demand End

It is worth noting that the chain of custody does not have to be initiated at the forest. Certain consumers have or are developing systems for assessing their suppliers and sources of forest products—in a sense working backwards through their supply chain. Some of these consumers are initiating such systems as part of their overall response to environmental issues; others are targeting wood products with a view to identifying those supplies that could eventually be certified, and/or to establish longer term, more stable supply sources. (See Box 6.3 and Box 6.4.) A questionnaire is often the starting point for gathering the relevant information, including information about regional sources, species, volumes, and the suppliers' knowledge of specific

---

Box 6.3

Responding to market pressures—Colonial Craft & B&Q

B&Q , Plc. is a "do it yourself" company based in England that has been involved in forest management certification for over four years, and is now working with all four of the first FSC-accredited certifiers: SGS-Forestry, Scientific Certification Systems, Soil Association, and Smart Wood. Colonial Craft Inc. is a molding and millwork company in Minnesota that was certified in 1995 by Smart Wood. In each of these cases, the companies made decisions to conduct thorough surveys of their current supply sources and, with strong internal staff support, began a process of gradual incorporation of certified forest products into their product lines. Both established target dates by which they hope to have all their products made of FSC-accredited certified wood, and both indicated a willingness to work with suppliers to meet strict chain-of-custody requirements. In each case, they have also exerted strong, but careful, pressure on their suppliers to move toward certification. They were motivated by a strong sense that consumer response would be positive and that purchasing certified products was "the right thing to do."

Box 6.4

Incorporating chain of custody into product design

Seven Islands Land Company was certified as a well-managed forestry operation in 1994 by Scientific Certification Systems Forest Conservation Program.
Seven Islands has historically focused on land management, having minimal
involvement in wood processing. However, based on innovative management
and creative marketing through a process called "out-sourcing," the company
has established relationships with a number of outside processing companies to
manufacture certified products ranging from golf tees to flooring. Seven Islands staff work closely with some processing companies in identifying markets
and ensuring timely delivery of high-quality products. Through such close relationships, Seven Islands and the intermediate processing companies are able
to incorporate chain-of-custody requirements from the initial design phase,
thus minimizing costs and maximizing profits.

source sites. Such a questionnaire is also typically used to evaluate the real-
ism of making the change to certified supply sources.

Chain-of-Custody
Constraints and Potential Solutions

A variety of additional factors may complicate full documentation of the
chain of custody. These include "gatewood," limited storage or separation
space, products from small landownerships, and multispecies, multicomponent products.

"Gatewood"

The term gatewood refers to wood that arrives at the doors or "gates" of a
manufacturing plant or mill with little or no previous planning and with little or no record as to the original source. Many mills rely on high amounts of
gatewood; it is not uncommon for pulp and paper mills or sawmills to rely on
gatewood for 100 percent of their supply. Obviously, gatewood presents a
major hurdle for certification.

In practice, it is often found that a high percentage of gatewood can be
tracked to its original forest source *if there is a reason to do so*. For example, International Paper Timberlands, a forest management subsidiary of International Paper, which manages roughly 6 million acres (2.5 million hectares) in

the United States, often sells pulpwood to independent contractors. These contractors may ultimately sell wood back to an International Paper mill. Until certification was considered, there was no reason to keep track of the wood source in this system.

Limited Space for Storage or Separation

If a company is committed to producing only certified forest products, there is no need to separate certified from noncertified product. However, for most of industry, this is not reality. Most companies start producing certified products as only part of their overall product line and, therefore, need to find ways of separating certified source material or final product from the noncertified. In many cases there just isn't enough space to do this. This is particularly true of sawmills or other processing facilities located in urban areas.

Options for solving space problems include establishment of intermediate yarding or collection areas, careful timing of harvest or collection for certified product so that it will run as a single batch, or single product runs through the processing facility.

Product from Small Landownerships

Many industries in the United States, Central and South America, Europe, and parts of Asia obtain raw material from numerous, geographically dispersed small forest blocks or ownerships. Often such wood comes in as gatewood. Harvesting on such properties may take place once every 10 or more years, there is often no long-term management contract, and in many cases there are absentee landowners living in far-off urban areas. Such ownerships represent an extremely important supply source for many industries, as in aggregate they often represent 50 percent or more of a region's woodlands (such as eastern United States and Costa Rica).

Numerous organizations are working on developing certifiable chain-of-custody systems for small landowners (the definition of "small" varies from place to place and is subject to interpretation—usually it means less than 50 hectares). In general, these organizations are either: (a) attempting to aggregate all products from the small ownerships at a central yarding area or concentration yard, documenting the source of wood, and subsequently putting aggregate volumes up for bid as a certified volume; or (b) establishing a close relationship with a mill accustomed to dealing with chain-of-custody issues.

In either case, landowners must be grouped under a common forest management framework that will ensure consistent forest management procedures. Landowner associations and cooperatives are two options. A third option is for a mill to develop a landowner assistance program that facilitates purchase of wood and consistent management quality.

Multispecies, Multicomponent Wood Products

Few forest products are made of only one species of wood. Many products include components that are solid wood, laminates or veneer, medium density fiberboard, or other forms of wood. For example, it is quite common for plywood to be tropical wood as a face veneer with temperate woods or reconstituted wood products as a core stock. For a product to be credibly certified, this means that all of the different components must be made of wood from certified sources. Though there are examples of certified multicomponent wood products (for instance, high quality "Ecopanels" produced by Buchner Veneer in California and marketed by Ecotimber International), such products are still few in number. The answer to such a challenge is to have more certified material available from which to build such products.

Chain of Custody—The Benefits

A number of approaches to chain-of-custody operation and audit have been cited above. In the context of certification systems, all examples are intended to verify that the wood product can be traced back to its source. Chain of custody is also valuable as a method of exerting greater management control over forests and manufacturing processes. Experience with certification, product tracking technology, and log exporting monitoring indicates that forest industry stakeholders can benefit from the implementation of paper, digital, or other chain-of-custody systems. The following benefits may accrue to parties implementing chain-of-custody systems.

First and foremost, chain of custody is a required element for those wishing to capitalize on the actual or potential markets for certified forest products. Thus, access to such markets should be mentioned as the first potential benefit.

Second, forest managers will have continuous access to information on forest product volumes and movements. This is important as the physical distance between logging areas and manufacturing increases. They will be able to assess the efficiency of product recovery and reduce wastage, reduce illegal harvesting, readily meet industry requirements and specifications, forge closer links with manufacturers and consumers, and provide clear and unambiguous information for investors. In countries where forest concession systems are being developed or redesigned (such as Guatemala, Honduras, and

Indonesia), the provision and use of better forest management information will be even more critical.

Third, manufacturers will have clearly presented information on the size, species, and quality of the raw material; form closer links with producers and consumers; provide the consumer with clear information on sources; and improve product quality and the ability to manage change. Improved materials and cost accounting is often an inherent product of chain-of-custody auditing. Wastage during harvesting or processing is a common source of lost revenue. Unless a company knows what is happening in terms of volume of product produced per volume of wood purchased, it is impossible to accurately quantify lost revenues. Implementation of chain-of-custody systems will provide important baseline and monitoring information to quantify these losses. For example, wastage due to inefficient systems during timber harvesting may be as high as 20 meters per hectare—material that could be processed and sold.

Fourth, wholesalers and retailers will have improved inventory control. In a well-designed system, they should be able to pinpoint areas of weakness in the supply chain, and forge closer links with producers and manufacturers, thereby reducing the number of middlemen and reducing costs. The drive to develop value-added manufacturing of wood products in producer countries has the potential of not only increasing revenues, but also of simplifying the supply chain. If wholesalers or retailers have the confidence that they can access a continuous supply of quality-assured products, they will be more likely to purchase directly from the producers. This is already happening—adding the additional assurance of material from well-managed sources (if this is specified) is not difficult.

Fifth, chain-of-custody certification is often a first step for a company that wishes to test certified markets. Thus it may be a way for the company to become publicly known as a company with an environmental commitment.

Sixth, for publicly owned or traded companies, the desire on the behalf of stakeholders, including consumers and regulators as well as other stakeholders, to verify environmental credentials is increasing. Having systems in place that demonstrate good management control over forestry and manufacturing operations may improve the image of the company to stockholders or the profile of a company as an investment opportunity. Shareholders will also have access to good independent information on which to base an investment decision.

Finally, the experience gained through private sector voluntary certification may also have benefits for the public sector. For example, a range of national governments worldwide already employ inspectors/certifiers to establish export and/or import monitoring systems for all goods. The governments of

Cameroon and Papua New Guinea, for example, have recognized that they are losing much potential revenue from logs that are illegally felled and removed from the country and are now in the process of establishing nationwide schemes to inspect all exports of logs and timber products, using private inspection or certification services. In all cases, of course, an inspection/certification agency charges for its services. However, the governments expect that increased revenues will significantly offset the cost.

Through chain-of-custody systems, governments will be seen as innovative in their approach to forest management policy. They will be better placed to strengthen their position within international fora. The forest industry will be a more attractive investment proposition, and stronger links between the producer and consumer countries will be promoted.

Forestry departments will be in a position to exert greater control over illegal logging. They will have a simple but effective method for monitoring log movements, obtain accurate and up-to-date information on royalty and levy incomes, and, in general, operate with better management information.

Conclusions

This paper provides a picture of what a chain of custody should look like and how it might operate. The key message is that the chain of custody is as important as the forest assessment component of the certification process. Indeed the chain of custody can have a life of its own beyond certification, because it represents good business management practice. Without a doubt, companies with sound procedures and practices will find it easier to satisfy chain-of-custody requirements within any potential certification system.

In addition, many of the chain-of-custody audit procedures are currently being used extensively in the global trade of manufactured goods. As those selling wood and wood products as well as the certifiers build on this experience and master the techniques to provide cost effective chain of custody, more accountable systems will exist for improved forest management in all parts of the world.

Part II

Key Issues

Forest product certification is a relatively new phenomenon. Many questions regarding the effectiveness, efficiency, and potential impact of certification remain. Part II seeks to raise some critical questions and highlight issues to consider as certification continues to develop and evolve.

Chapter 7 examines the impact of certification as a policy instrument. Environmental marketing, characteristics of an effective certification system, and preconditions and limitations are outlined in this chapter. Chapter 8 explores the match between the objectives of many environmental organizations and the aims of certification. Included in this chapter is a description of some of the actions nongovernmental organizations might take to support certification, ranging from policy activism and lobbying to consumer education and business partnerships.

Chapter 9 discusses the possible impact of certification as a catalyst for change. Possible impacts include financial incentives, increased market potential, and model forestry initiatives. The economic aspects of certification are further described in Chapter 10. This chapter includes some of the direct and indirect costs of conducting the certification inspection and monitoring the chain of custody. Also included are some of the economic aspects.

Chapter 11 explores some of the potential negative and unintended consequences of certification. Barriers to trade, inequitable access to developing countries, bias against small enterprises, and forest product substitution are some of the concerns raised.

Most of the discussions in this book have centered on the certification of timber. Chapter 12 includes a discussion of certification of nontimber forest products (NTFPs). This chapter describes the importance of NTFPs to sustainable forest management systems, provides examples of NTFP management systems, and raises a number of questions and issues to consider in the certification of these products.

Finally, Chapter 13 describes some of the research gaps and information needed to support forest product certification. New timber harvesting methods, the silvicultural effects on commercial and noncommercial species, ecosystem processes, and complex social systems all contain areas for further research.

Chapter 7

Certification as a Policy Instrument

Chris Elliott

New Actors and Mechanisms in Environmental Policy

Traditionally, in most countries policies concerning forestry or the environment are made by government. The role of NGOs and the private sector has been to seek to influence the formulation and implementation of policy by government. More recently, there appears to be a tendency for the private sector and NGOs to look for policy instruments that permit them to have a stronger role in development and implementation. This sometimes involves collaboration between NGOs and the private sector, with the government taking a monitoring and supporting role.

The reasons for this evolution, currently limited to parts of Europe and North America, are perhaps to be found in both the dissatisfaction of NGOs with their lesser role and a tendency toward the reduction of the role of government in the economy under the influence of neoliberal economic theories.

In the private sector this development in the environmental field follows a similar path to what happened in the social field, particularly in the United States: Large corporations developed their own policies on employment of minorities, investments in South Africa, and so on, beginning in the 1970s (Pattie 1994).

A further trend in the same countries is for individuals who are interested in particular environmental or social issues to modify their consumption patterns to promote certain objectives (such as boycotting tropical timber and using public transport). This has been actively encouraged by some NGOs.

Timber certification has developed in this context. There are two components of timber certification. The first is certification of forest management, also referred to as forest auditing. This involves inspection of forest management on the ground against specified standards and review of documents such as management plans, inventories, and so on. Certification of forest management can be done at different levels (forest management unit, forest owner, region, or country). Existing certification programs work at the level of the management unit.

The second component is product certification. For a consumer's purchasing choice to be influenced by certification the certified timber must be followed through the production process to the product the consumer purchases, whether it be a table or a plank. The chain of custody involving log transport and processing, shipping, and further processing is also subject to certification. (See Chapter 6.)

As mentioned earlier, the intended mechanism of timber certification is to link the "green consumer" who is allegedly willing to pay more for sustainably produced wood and wood products to producers who are seeking to improve their forest management practices and obtain better market access and higher revenue. Certification is thus a category of environmental labeling that, according to the U.S. Environmental Protection Agency, "strive(s) to make credible unbiased and independent judgments in certifying a claim or product" (EPA 1993).

In a recent report the EPA observed that environmental labeling programs are increasingly used (by NGOs and governments) as "soft policy tools" harnessing market forces to reach certain environmental goals. The EPA reported that programs are underway in 21 countries and noted that "several surveys indicate that a majority of Americans consider themselves to be environmentalists and would prefer to buy products with a lessened environmental impact when quality and cost are comparable" (EPA 1993).

Timber certification was originally promoted by conservation NGOs such as Friends of the Earth and WWF (Elliott 1994). More recently governments and the timber industry have begun to express an interest. The establishment of "labeling and tracking systems to allow consumers to assess the sustainability of timber sources" was recommended in *Caring for the Earth*, published by WWF, IUCN, and UNEP (Anon. 1991). In a 1993 publication entitled "Surviving the Cut" the World Resources Institute made a positive assessment of the possibilities of timber certification and noted that "timber certification systems offer a first critical step toward sustainable forestry" (Johnson and Cabarle 1993). IIED, The London Environmental Economics Center, recommended certification as an incentive for sustainable forest management in a recent report (LEEC 1993).

World Bank forest policy states (1991) "experience with other products suggests that consumers will modify their behavior substantially if they are given information on the ecological sustainability of the production process. For this reason the international community should encourage organizations like the International Tropical Timber Organization to develop programs of green labeling to permit preferential market treatment for wood grown under sustainable circumstances. In addition to lowering the overall demand for wood produced by unsustainable practices, such a scheme would remove the disincentive for adopting improved management practices that might otherwise diminish competitiveness." In this chapter, certification is first examined from the perspective of "environmental marketing" and then from the "policy instrument" perspective.

Objectives of Timber Certification

In discussions on timber certification two objectives are usually identified: (1) to improve forest management, and (2) to ensure market access for certified timber. In the next section, timber certification is discussed in the context of environmental marketing, in line with the second objective. Following this, the joint management objectives are dealt with when certification is reviewed as an instrument of forest policy.

Conservation NGOs tend to focus on the first objective and retailers and timber traders on the second. It is a characteristic of timber certification that it is being promoted by these two different groups of people with somewhat different objectives. There is a potential problem in that the achievement of one of the objectives will not necessarily mean that the other is also achieved. However, certification is not likely to be successful in the long term unless both are achieved. One of the purposes of this study will be to clarify the perceptions of different actors on the objectives of certification.

In addition to the primary objectives of certification, a number of secondary objectives can be identified (Simula 1993):

- Improved transparency and control of forest management, particularly with respect to illegal logging activities (this seems to be of particular interest to the Indonesian authorities in their deliberations on certification)
- Higher recovery of royalties, forest taxes, and other fees

- Increased availability of funds for forest management
- Internalization of environmental costs in the production costs of timber and timber products
- Improved total productivity and cost savings in the production chain from forest to final user, with potential reduction in the number of intermediaries

Timber Certification and Environmental Marketing

As mentioned above, timber certification is a form of environmental labeling, which is itself a category of environmental marketing. A recent EPA report notes that following increasing news coverage of environmental issues in the 1980s, the U.S. marketplace has seen significant increase in the number of new products sold making environmental claims. Two broad categories of such claims can be recognized. These include first-party environmental marketing in which the manufacturer or retailer makes a claim about the environmental performance of the product (e.g., recyclable) and third-party environmental labeling in which an independent third party makes a credible, unbiased judgment in certifying a claim or product. Such programs are intended to provide consumers with information or assessments that are not often apparent or available to the consumer, helping the consumer to make purchasing decisions based on the environmental impacts of products (EPA 1993).

Programs may be mandatory or voluntary. Mandatory programs are required by laws or regulations and consist of either hazard warnings such as those found on cigarette packages in some countries, or information disclosure labels revealing important facts about a product (such as fuel consumption of a car).

There are three kinds of voluntary programs. These include eco-labels, voluntary certification, and report cards. Eco-labels (also referred to as multi-issue labels or Seal-of-Approval) identify products or services as being less harmful to the environment than other similar products or services, through the use of a recognizable stamp or label. Most programs attempt to base themselves on life-cycle analysis (LCA), a complex analytical process still under development, which attempts to identify and measure the full environmental effects associated with the product's life cycle—extraction and processing of raw materials, manufacturing, transportation, distribution, use, recycling, and disposal (OECD 1991).

In voluntary certification (sometimes referred to as single-issue label), most current labels, including timber certification, confine themselves to the part of the life cycle of the product considered to be of concern to consumers.

Thus, for example, timber certification concentrates on the sustainable management of the forests that produce the timber and not on emissions during transport and processing, the environmental impact of wood preservatives, or other issues. Like eco-labels, certificates convey a positive message to consumers in that they identify products to which environmentally concerned consumers will hopefully be attracted.

Interestingly, the timber certification bodies currently active are all independent of government (they are run by NGOs or for-profit companies), whereas most of the other environmental certification programs running in 21 countries have some form of government involvement.

Report cards seek to offer consumers neutral information (rather than positive information as with eco-labels and certificates) about a product or service. Report cards are generally based on some form of LCA, but contrary to eco-labels, the data are presented in some detail (energy consumption, various emissions, and so on) allowing consumers to weigh for themselves what the most important environmental impacts are.

Timber Certification as a Policy Instrument

Environmental labeling in general, and timber certification in particular, have been classified as combined instruments of environmental and trade policy. Baharuddin and Simula (ITTO 1994) describe the genesis of environmental labeling as follows:

> The current extensive public concern given to environmental labeling can be interpreted as a sign of growing concern worldwide on sustainability of development and, in particular, trade's impact on the environment. The interface between the two is characterized by the possible role of trade policy instruments in the pursuit of environmental objectives, and the impact of environmental policies on trade. Environmental labeling can be seen as a combined instrument with clearly defined environmental objectives which are to be achieved by trade measures.

The literature on policy instruments has so far not accorded much attention to certification. This may be partly because most of the literature concentrates on the policy tools governments use, and governments have not used certification so far. As noted above, the currently active certifiers are NGOs or private companies. There has been no systematic study of policy instruments used by conservation NGOs.

Despite this, some authors have explicitly recognized certification as an instrument of policy (Linder and Peters 1989). There are various classifications of policy instruments. Perhaps the most useful for the purposes of this study divides instruments into five broad categories: authority, incentives, capacity building, symbolic and hortatory, and learning (Schneider and Ingram 1990). The authors of this classification provide an interesting analysis of each category and the behavioral assumptions underpinning it. They note that if people are not taking actions needed to ameliorate social, economic, or political problems, there are five reasons that can be addressed by policy. Two of these reasons are of direct relevance to timber certification: People may lack the incentives or capacity to take the actions needed, and it is unclear what individuals should do or how they might be motivated.

In the first case the appropriate policy tool would come from the category of incentives—in the second, from learning. Schneider and Ingram (1990) note: "Incentive tools assume that individuals are utility maximizers who will change their behavior in accord with changes in the net tangible payoffs offered by the situation. Learning tools assume agents and targets do not know what needs to be done, or what is possible to do, and that policy tools should be used to promote learning, consensus building, and to lay the foundation for improved policy." For the purposes of this study timber certification will be considered to fall into the category of incentive tools, although it also includes some elements of a learning tool.

Characteristics of an Effective Certification System

There are several political, legal, and operational characteristics of an effective certification system. To be internationally acceptable, timber certification systems must be voluntary and must apply to all types of timber and non-timber products. (See also Chapter 11 and Part III, "Geographic Regions.")

The following operational characteristics were mentioned earlier in relation to effective certification systems:

- Credible to consumers and conservation NGOs
- Objective and measurable criteria
- Reliable and independent assessment
- Independence from parties with vested interests
- Cost-effective
- Transparent to allow external judgment
- Institutionally and politically adapted to local conditions
- Goal orientated and effective in reaching objectives
- Acceptable to all involved parties
- Use of national-level forestry criteria compatible with generally accepted international principles

Mechanisms for harmonization and mutual recognition of different timber certification systems are needed if confusion is not to arise as the number of schemes increases.

Preconditions for Certification to Be an Effective Policy Instrument

If timber certification is to be effective in reaching the objective of improving forest management, a number of preconditions must be satisfied because certification alone cannot achieve this objective. As Baharuddin and Simula note (ITTO 1994), "the effectiveness of certification as a policy instrument itself should in principle be assessed in the context of a policy package where certification is a component."

Similarly a recent World Bank paper (Crossley et al. 1994) notes that

> Enabling producers to capture greater revenues [through certification] will not ensure that improved forest management systems and deforestation rates ensue. The policy environment for most [tropical] producers is highly distorted. Market and policy failures in the forest and related sectors are, in fact, the primary cause of deforestation and forest degradation. These failures must be addressed concurrently with the introduction of a certification system to enable the market signals it creates to effectively and properly influence forestry in developing countries.

Policy Preconditions

Although timber certification affects only forests destined for timber production (known as the Permanent Forest Estate in some countries), sustainable forest management can be achieved only if national and international policies identify sustainable management as a priority across all relevant sectors. A recent review of this topic (de Montalembert and Schmithüsen 1993) notes that today forests must be managed in a much more complex and interdependent context than in the past, where timber production was the main objective in most areas. This requires a partnership among major actors and beneficiaries based on a recognition of the diversity of interests present. The authors consider that the following aspects of the policy framework are particularly relevant:

> *Land use policies and planning.* The importance of the forest resources of a country must be recognized and land-use planning must make provision for the maintenance and sustainable management of forests.

Macroeconomic policies and structural adjustment. Attention must be given to the potential negative impacts on forests of macroeconomic policies, including privatization and debt-reduction schemes.

Policy interactions. Interactions among related sectors such as agriculture, industry, and energy must be considered, and impacts in areas bordering forests must be monitored.

Conservation and wise utilization of forests. Conservation of forests must obtain clear recognition and support as a national priority in forest, environmental, and development policy.

The behavior of various social groups. Social groups having an interest in forests should be a focus of forest policy. Forest policy must therefore recognize different user-groups and encourage flexible approaches to take their interests into account.

Forest dwellers and neighboring communities. Peoples who are often economically marginalized must be involved as preferential partners and beneficiaries.

Fiscal policies. Fiscal policies should not only be judged in terms of revenues accruing to government through fees and taxation, but by the way in which they influence the willingness and ability of the entity in charge of forest management to invest in its sustainability.

Price policies. Price policies should encourage prices of forest products to reflect real economic values effectively while monitoring substitution with other materials. Forest product markets should ensure a fair share of revenues to those who are actually responsible for managing the forests.

Turning to policies for managing the Permanent Forest Estate (forests set aside for timber production) on a sustainable basis, the preconditions have been identified as follows (IIED 1988): (1) government resolve; (2) a sound political case for the selection of a Permanent Forest Estate; (3) long-term security for the forest estate, once chosen; (4) a market for forest produce; (5) adequate information for the selection of the forest estate and for planning and controlling its management; (6) a flexible predictive system for planning and control; (7) the resources needed for control; and (8) the will needed by all concerned for effective control.

Socioeconomic Preconditions

Certification will achieve its objectives only if the benefits of certification exceed the costs for the key actors in the process. It is currently difficult to make a comprehensive assessment of the costs and benefits of certification.

Costs can be divided into three categories: (1) the cost of improving forest management at the level of the management unit so that it reaches a standard that is acceptable to certifiers; (2) the cost of certification of forest management (the forest audit); and (3) the cost of product certification (chain of custody). The benefits of certification may be divided into both market benefits and nonmarket benefits.

The potential market benefits are market access and pricing (the "green premium"). The former is already a reality in some market sectors (for example, do-it-yourself chains in the United Kingdom) or countries (such as Holland) that are committed to purchasing or importing only sustainably produced timber by 2000 and are seeking certification as a means to achieve this. Market studies in the United States and the United Kingdom have shown an interest in some sectors of the market for paying a "green premium" for certified timber, but Baharuddin and Simula (ITTO 1994) conclude that "there is not yet convincing evidence of an existing price premium for sustainably produced, certified timber and timber products on the market . . . there are some market segments, however, where willingness to pay a premium can be observed and could be exploited by trade." These issues are discussed further in Chapter 10.

Conclusions

Certification can best be understood as a forest policy instrument intended to provide market incentives. Like any policy instrument it has limitations, and a number of conditions are required for it to be effective. If these points are kept in mind, certification probably has a promising future.

References

Anon. 1991. *Caring for the Earth; a Strategy for Sustainable Living.* IUCN, WWF, UNEP, Gland. 227 pp.

Crossley, R., C. A. Primo Braga, and P. Varangis. 1994. *Is There a Case for Tropical Timber Certification?* The World Bank. Washington, D.C. 33 pp.

de Montalembert, M., and F. Schmithüsen. 1993. Policy and Legal Aspects of Sustainable Forest Management. *Unasylva* vol. 4c, no. 175.

Dudley, N. 1992. *Forests in Trouble; a Review of the Status of Temperate Forests Worldwide.* WWF International, Gland. 260 pp.

Elliott, C. 1994. Timber Certification and The Forest Stewardship Council. Paper presented at the Malaysian Timber Industry Development Council Seminar on "Trade of Timber from Sustainably Managed Forests." April 1994, Kuala Lumpur.

Environmental Protection Agency. 1993. *Status Report on the Use of Eco-Labels Worldwide*. Washington, D.C. 215 pp.

ESE. 1993. Study on a Certification System for Sustainably Produced Timber (Executive Report). Environmental Strategies Europe, Brussels. 20 pp.

IIED. (1988) Natural Forest Management and Timber Production. Pre-Project report to the International Tropical Timber Organization. International Institute for Environment and Development, London. 163 pp.

ITTO. 1994. Report of the Working Party on Certification of all Timber and Timber Products. ITTO, Yokohama, Japan. 200 pp.

Johnson, N., and B. Cabarle. 1993. *Surviving the Cut, Natural Forest Management in the Humid Tropic*. World Resources Institute, Washington, D.C. 71 pp

LEEC (London Environmental Economics Center). 1993. The Economic Linkages between the International Trade in Tropical Timber and the Sustainable Management of Tropical Forests. Report to ITTO. LEEC, London. 132 pp.

Linder, S. H., and B. G. Peters. 1989. Instruments of Government: Perceptions and Contexts. *Journal of Public Policy* vol. 9, no. 1, 35–58.

OECD (Organisation for Economic Cooperation and Development). 1991. "Environmental Labelling in OECD Countries." OECD, Paris, France.

Pattie, D. 1994. *Environmentally Based Marketing Programs: a Strategic Approach to Enhancing Marketing Performance*. ITTO, Yokohama, Japan. 51 pp.

Schneider, A., and H. Ingram. 1990. Behavioral Assumptions of Policy Tools. *Journal of Politics* vol. 52, no. 2, 510–529.

Simula, M. 1993. Consumer Initiatives on Eco-labeling of Tropical Timber and Their Implications for Indonesia. Seminar Sehari Eco-labeling Produk Hasil Hutan. Jakarta 13 September 1993.

The World Bank. 1991. A World Bank Policy Paper, The Forest Sector. The World Bank, Washington, D.C. 98 pp.

Chapter 8

Role of NGOs

Richard Z. Donovan

Certification raises important questions as to the roles that NGOs should play in the pursuit of sustainable forest management. In spite of certification's relatively short history, some preliminary implications can be drawn regarding the different roles NGOs might play, from ethical, management, educational, or financial perspectives.

Many of the people managing nonprofit organizations are beginning to face decisions about whether to be involved in forest management certification as direct participants, distant or close observers, educators, or lobbyists. Experience in other sectors (such as organic food certification) indicates that nonprofits could fundamentally influence the impact of certification on commercial forestry.

Historical Perspective

Nongovernmental, nonprofit organizations have played crucial roles in raising the visibility of independent forest management certification. The concept of purchasing environmentally or socially sensitive wood products was first introduced by Friends of the Earth, U.K., through the publication of the first "Good Wood Guide" in 1988 (Friends of the Earth 1990). This guide offered consumers guidance for purchasing wood products that were less environmentally damaging to tropical forests. In 1990, certification was launched when the U.S.-based Rainforest Alliance's Smart Wood Program certified Perum Perhutani in Indonesia as the first independently certified "well-managed" source of tropical hardwoods. At the same time, in response to consumer interest and lack of action by the International Tropical Timber Organization (ITTO), World Wide Fund for Nature (WWF) United Kingdom launched the "1995 Group." This initiative, which included major retail and wholesale companies in the United Kingdom, was designed to reduce consumption of unsustainably produced forest products and increase the use of wood from "well-managed" forestry operations. The 1995 Group initiative in the United Kingdom explicitly sought to capitalize on consumer concern to create market forces (purchasing policies) that would favor wood purchases from operations committed to sustainable forest management.

In these cases, nongovernmental organizations played different roles. In most cases the NGOs attempted to work with supportive elements in business. NGOs established relationships with companies wishing to purchase certified forest products. For Rainforest Alliance, Smith & Hawken, Inc., a mail-order garden equipment and furniture company in California, was looking for alternative sources of teak, which at the time came from questionable forest management operations in Burma. Rainforest Alliance's certification of Perum Perhutani, designed to provide an alternative to purchases from Burma, was closely followed by for-profit Scientific Certification Systems' (SCS) and Rainforest Alliance's certification of the Plan Piloto Forestal in Mexico. Similarly, buyer interest occurred in the Knoll Group, an office furniture manufacturer that supported SCS certification work in Mexico. Subsequently, assessments led toward certification of Menominee Tribal Enterprises in Wisconsin and the development of a line of office furniture products from well-managed forests.

As these initial examples show, some NGOs have already played important roles in certification, working in close contact with the business sector. Certification's increased visibility in the marketplace is putting many NGOs in a challenging situation. Certification implies acceptance of the need for timber harvesting. It also requires direct working with for-profit companies. As such, it is forcing many NGOs to reevaluate their policies, positions, or programs related to commercial forestry, timber harvesting, tree plantations, and certification itself.

Consistent NGO Themes

NGOs have historically focused on a core of values related to sustainable forest management, including (1) environmentally and ecologically sound forest practices, (2) positive socioeconomic benefits for local people, and (3) sustained-yield management or production of forest products.

Though NGOs have placed a priority on these values, they have had to push strongest to get ecological and, where there are indigenous peoples, socioeconomic issues on the agenda in commercial forestry.

As NGOs become involved in certification, they are often forced to make important decisions on what their "position" is regarding particular issues, such as harvesting in old-growth forests or the use of tree plantations, clearcutting, or chemicals. Not all issues apply in all regions, but increasingly these issues are forcing NGOs to become more educated about the practical aspects of each issue and commercial forestry. Many NGOs choose to not take positions—others do. Whether as a policy advocate or as a forest management practitioner, in-depth familiarity with on-the-ground forestry techniques is becoming an increasingly important factor in determining NGO forest policies. Involvement in certification facilitates such understanding; it also appears to change an NGO's perspective to less rigid, more practical perspectives.

Ecological Forestry

For many years, some NGOs have been on an almost crusadelike effort to incorporate ecological values into commercial forest management. The key objectives of this effort are:

- Protection of threatened and endangered species
- Protection of old-age, or "ancient," forest ecosystems
- Elimination of clearcuts, or at a minimum a reduction in their size and frequency, usually implying reduced use of "even age" (versus uneven age or selection) forest management techniques
- Incorporation of broader landscape-level ecological and biodiversity concerns into forest management planning, including management of game (hunted) and nongame (not hunted) wildlife species
- Protection of watersheds and aquatic ecosystems (estuarine, riverine, and oceanic)
- Use of reforestation for restoration ecology, including expanded use of native tree species in reforestation programs and control of exotic tree species
- Minimization, if not complete elimination, of use of chemicals

Positive Impact on Local Communities

Concern about local communities, indigenous peoples, and worker conditions has also been a high priority. Environmental NGOs have teamed up with grassroots rural development and human rights groups to stress that commercial forestry should have a positive impact on both modern and traditional communities. One of the most visible issues has been the desire to see government and industry organizations legitimize or, at a minimum, pay more respect to the customary resource management rights and practices of indigenous communities in countries such as Indonesia, Malaysia, Brazil, Canada, the United States, Ecuador, Peru, Guyana, and Surinam.

Environmental NGOs maintain that there are intrinsic links between ecological sustainability and socioeconomic stability at local and regional levels. To environmentalists, these concepts are inseparable. In practice, this philosophy requires that they work with local people and communities to play an active role in commercial forestry by helping to define the mechanics of, and limits to, timber harvests on both public and private lands.

From a conservation perspective, one only has to look at the conflict created through "boom and bust" cycles of logging activities in regions such as Southeast Asia (Thailand, Philippines, Malaysia, and Indonesia), North Central United States (Minnesota and Michigan), and the Pacific Northwest of Canada and the U.S. to find examples of what happens when the demand for wood volume exceeds sustainable productive levels. Though the definitive "sustainable" levels of timber production in a region may be difficult to define—often too little is known about ecological, silvicultural, or socioeconomic dynamics to definitively establish them—environmentalists believe that clearly unsustainable levels can be identified, given an analysis of existing environmental and socioeconomic impacts and conflicts that are occurring.

To some (though by no means all) in forest industry, environmentalist concerns about community issues, worker conditions, or the rights of indigenous people smack of idealistic "social engineering," or as an impractical refutation of the modern nation-state, or capitalism.

Sustained Yield Forestry

Sustained yield forestry focuses principally on ensuring that the annual timber harvest volume does not exceed the volume of annual tree growth. This concept originated in Germany in the 19th century and is embraced by most in forest industry. However, as defined by environmental NGOs, sustained yield forestry should be based on natural ecological limitations and forest potential. In other words, sustained yield forestry that achieves a balance between growth and harvest should not do so at the cost of ecological integrity. Typical issues that come up in this context are forest practices that are perceived to be threatening to the integrity of the ecosystem—for example, large clearcuts or widespread use of chemicals.

Roles for NGOs

In certification, NGOs have sometimes played roles that might be regarded as traditional, such as lobbying, education, and acting as environmental "watchdogs." But they have also begun to play some relatively new or unfamiliar roles, requiring in many cases different skills or perspectives. This chapter examines the following potential NGO roles in certification: (1) general public education; (2) policy activism, including participation in standards-setting forums or committees for certification by creating a policy environment that favors sustainable forest management and, either directly or indirectly, certification by participating in the growing "certification industry" in general, including certifiers and international certifier accreditation agencies (such as the Forest Stewardship Council or FSC); (3) acting as watchdogs in terms of market claims of sustainability by forest product companies; (4) working in partnerships with industry to help companies meet certification criteria or improve the overall sustainability of commercial forestry operations; (5) opening up the marketplace for certified forest products by educating the public or lobbying companies, public agencies, or their peers in the conservation community; and (6) actually implementing certification programs.

Role #1—Public Education and Communication

There are many different ways to carry out public education activities. In the 1970s and 1980s, one tactic of choice for public education was the boycott. In the case of forests, it was the tropical timber boycott, chosen as a tool to bring public attention to the role of forest industry and wood products consumers in tropical deforestation.

The first NGOs to support certification (such as Rainforest Alliance and WWF) were not convinced that timber boycotts would solve tropical deforestation. Though they did not outwardly oppose the tropical timber boycotts proposed by other NGOs, they did seek other options because they felt boycotts were too simplistic. In general their perspective was that wood was needed and timber was going to be cut. For them, the question was not if timber would be cut, but how.

Thus, they began to work on certification in the late 1980s. With roughly 10 years of experience in organic food certification to look at, these NGOs began to design procedures for forest management certification. In a relatively short time—two to three years—positive consumer response and retail company receptivity to tropical forest-oriented certification, and increasing visibility of forestry problems in other regions (for instance, temperate rainforest and boreal forest) resulted in expansion of the certification concept to cover all forests.

Initially, certification provided NGOs with an alternative to tropical timber bans and boycotts—a positive, market-oriented incentive for commercial forest industries. However, the intrinsic logic of certification—independent recognition of good forest managers rather than blind acceptance of industry claims or environmentalist condemnation—resulted in certification beginning to gain force as a tool for public education about forest management. This change represents a fairly fundamental shift, with many NGOs beginning to focus on how forest industry would be structured and with what types of protection for biodiversity, indigenous people, and local communities.

Since roughly 1992, public education regarding certified forest products has begun to take place in northern Europe, the United States, and Costa Rica. Notably absent are major public education efforts in Canada (a huge producer) and Japan (a huge consumer). Though these efforts are a beginning, and the next five years will certainly see major, more sophisticated efforts, there are at least a few experiences where NGOs have attempted broad-based education with either consumers, companies, government, and/or schools.

In the Netherlands, Hartvourhout has gradually increased public awareness about the availability of certified forest products and the importance of Dutch companies buying them. In Costa Rica, the USAID-funded REFORMA project has provided financial support for the Conservation Media Center of the Rainforest Alliance in the development of a broad-based public education program focusing on sustainable forest management and certification. Activities include newsletters, posters, press releases, sustainable forest management field days, and other media events. WWF, through the 1995 Group in the United Kingdom, has gradually increased public information to consumers as their initiative has gained force.

The lack of supply of large amounts of certified wood products has slowed some organizations' public education efforts. Aggressive development of public education programs will probably now begin, as the concept is more widely accepted among forest industry and the supply of certified wood products increases.

Role #2—Policy Activism

Policy activism is one of the most traditional NGO roles. In this role, conservation or environmental NGOs have largely put forest management certification on the global forest policy agenda. As discussed elsewhere in this volume, conservation organizations formed a critical supportive core for the establishment of the Forest Stewardship Council. Three levels of NGOs have historically been involved. These include (1) international NGOs—WWF network (International, United Kingdom, U.S.A., Canada, Brazil, Indonesia, and Malaysia), Cultural Survival, World Resources Institute, World Rainforest Movement, Greenpeace, Rainforest Action Network (RAN); (2) national NGOs—Fundacion Natura of Ecuador, National Wildlife Federation of the United States, Sierra Club, Friends of the Earth United States and United Kingdom; and (3) regional or community-based NGOs—SOS Atlantica of southern Brazil, Institute for Sustainable Forestry in California, Rogue Institute for Ecology and Economy in Oregon, and Silva Forest Foundation in British Columbia, Canada.

National and international NGOs have tended to focus on either environmental or indigenous peoples' issues. Local NGOs have focused on these topics as well, but they have tended to press harder on general socioeconomic issues, due to the stronger impact of forest industry on rural, rather than urban, economies.

NGO policy activism has largely focused on participating in numerous forums with the intent of establishing definitions, principles, criteria, or indicators for sustainable forest management at international, national, or regional (subnational) levels. Because NGOs have a relatively high level of credibility with consumers (they are usually perceived to have less of an inherent conflict of interest), their involvement is often an indicator of, and a strong influence on, the success of the process. A short list of activities and fora in which major international NGOs have been active in this regard includes:

- Forest Stewardship Council (Principles and Criteria for Forest Management)
- Helsinki Process (Criteria and Indicators for Biodiversity Conservation and Sustainable Forest Management)
- Commission for Sustainable Development
- Inter-Governmental Forestry Working Group (Malaysia–Canada Initiative)
- Montreal Protocol—Criteria and Indicators for Temperate and Boreal Forests
- International Tropical Timber Organization (Criteria and Indicators for Sustainable Tropical Forest Management in Natural Forests and Forest Plantations; Guidelines for Managing Biodiversity in Production Forests)
- IUCN Species Survival Commission (Guidelines for the Sustainable Use of Wild Species by the Sustainable Use of Wild Species Specialist Group)
- International Organization for Standardization (ISO), which is developing standards for environmental management, potentially including forestry
- FAO, which has produced a "Code of Forest Practices"

As consumer interest in sustainable forest products has grown, there has been an increasing intensity of discussion to define sustainable forest management in commercially oriented forest products and industry settings. Usually these discussions focus on developing either product or manufacturing process standards, specifications, and in a few cases principles, criteria, or indicators for sustainable forest management. Two examples are the ISO process and the American Society for Testing and Materials (ASTM).

The Canadian Standards Association (CSA) has a joint effort with the Canadian Pulp and Paper Association (CPPA) to define standards for sustainable forest management, with the ultimate objective of using such standards to certify under the ISO umbrella. Conservation NGOs were invited to participate in the process but, as of early 1996, most have withdrawn from the process because they felt their input was not having much impact. They believed the process to be too closed, and the process appeared to be increasingly oriented toward sustained yield forestry, with little or no emphasis on ecological or socioeconomic issues. In this case, the departure of many key NGOs has threatened the credibility of the process, though most of Canadian forest industry appears to remain committed to the process.

In the United States, the ASTM has established hundreds of standards for various commercial products, from concrete to electrical appliances. ASTM members include manufacturing companies, architects, engineers, procurement agencies, and industrial designers. The publication of an ASTM standard can have major implications for any product because these standards are often used as guidelines for procurement policies of large public sector agencies or as the basis for specifications established by designers or architects for large-scale construction projects (such as office buildings). In late 1994, at the request of some ASTM members, a task force was established to begin developing standards for "sustainable wood." Initial drafts were developed, mostly by architects, based principally on information obtained through the Forest Stewardship Council founding members in the United States. However, when large forest product companies (such as Weyerhauser and Georgia Pacific) and industry associations (for example, American Forests and Paper Association or AFPA) found out more about the ASTM sustainable wood standard initiative in early 1995, they became very proactive in the process, seeking to have more participants on the task force. In response, conservation NGOs are also attempting to strengthen their hand in the process. The ultimate product of this process is difficult to predict.

Historically, conservation NGOs have been quite active in the national and international processes such as the International Tropical Timber Organization, the Montreal Protocol, or the Helsinki process, but they have been much less visible in standard-setting discussions at organizations like ISO or ASTM. The increasing visibility of voluntary mechanisms for environmental change, such as certification or labeling, has forced NGOs to become more active. Participation in these efforts requires major inputs of time to first understand how such standard-setting institutions work and then to influence them. Currently, the Global Forest Policy Project (supported by National Wildlife Federation, Sierra Club, and Friends of the Earth in the United States), WWF, Greenpeace, and various other NGOs are attempting to closely track ISO processes.

As certification has gained strength, a number of industry-supported initiatives have occurred in various countries. Some are clearly certification-oriented, such as the Camara Nacional Forestal initiative in Costa Rica, or the Indonesian Forest Concession Holders Association (or APHI) initiative. In these cases, industry has accepted the idea of certification and independent audits, but is attempting to control the process of certification standards development.

Other recent initiatives are supposedly not certification, but perhaps have been designed to respond to the increasing force of independent certification; an example is the American Forests and Paper Association Sustainable Forestry Initiative (Wallinger 1995). In this case, forest industry in the United States has designed a program that establishes standards for industry improvement in environmental practices (through "best management practices") but that is largely based on company self-evaluation, excludes social issues, and has only a limited amount of interaction with environmental groups in the standards development process. The program does, in theory, represent an opportunity for positive change. In the United States, AFPA is now aggressively marketing their initiative as an example of how industry takes care of the forest. This strategy is an alternative approach to gaining consumer confidence about industry's environmental practices. However, the extent to which this approach will have credibility and reduce public concern remains to be seen.

All the initiatives above require policy activism by NGOs—all represent challenges. Often the decision of a particular NGO to participate has crucial implications for the future of an initiative. One NGO's desire to work with an industry initiative, such as the AFPA's initiative, because they see a potential for positive internal change in industry, may undermine another NGO's independent certification initiative. This doesn't imply that participation would be inappropriate—only that NGOs must be careful to plan their involvement strategically and be clear about the conditions and potential implications of each role.

Role #3—Sustainability Claims Watchdog

Flora and Fauna International in the United Kingdom has investigated claims made by industry about sustainability in public advertisements, catalogues and promotional literature. To do so, they requested copy of advertising or promotional materials, wrote to each company asking the basis for each claim, evaluated the claim and, where appropriate, brought spurious claims to the attention of a British government office focused on ethical trade practices.

In the United States, the Council on Economic Priorities produces independent reviews of the environmental practices of many companies in different sectors, including forest products. Separate public reports on over 20 major forest products companies are available through the council, reviewing industry actions related to environmental protection.

These types of review activities, while often subjected to harsh industry criticism, provide a valuable service to the public. They also provide an interesting complement to certification. Companies subjected to negative reviews by such efforts would seem to have the most to gain from certification. Given the proliferation of advertising claims about sustainability that are made by forest industry, this would also seem to be fertile ground for NGO activity in most countries that are major producers or consumers of forest products.

Role #4—Forest Management Partnerships

A number of NGOs are working in close partnerships with for-profit companies to establish sustainable forest management model projects or to encourage more sustainable practices within those companies. Such efforts often involve large companies and biological conservation. Georgia Pacific, Inc. and The Nature Conservancy are currently cooperatively managing a 20,000-acre block of coastal floodplain forest in South Carolina with an emphasis on protection of biological conservation corridors as well as the use of environmentally sensitive harvesting techniques (for example, helicopter logging). The Nature Conservancy is also involved in similar efforts in Belize (with Program for Belize), Paraguay (with Fundacion Moises Bertoni), Wisconsin (Baraboo Hills Project), and Virginia (Clinch River Reserve Project), where sustainable forest management and economic development objectives are combined with biological conservation. In the United States, the Conservation Fund is working with International Paper on a similar effort, with an emphasis on establishing conservation zones and restoration ecology with a native timber species, longleaf pine. In each case, certifiable forest management may be one outcome of the partnership. However, in most cases certification was not an original objective. Also, there is a danger that such efforts by large forestry companies may deflect concern from environmentalists or government. Such efforts are unlikely to be successful over the long term unless they represent part of a broader commitment to positive environmental change throughout a particular company.

Estudios Rurales y Asesoria in Mexico has been working in partnership with a number of community-owned forestry operations ("ejidos"), assisting them not only in biological conservation issues, but also in forest management planning, timber production, community involvement, and wood processing. Through the USAID-funded SUBIR project in Ecuador, the Jatun Sacha Foundation is providing technical assistance and training to conduct forest inventories and develop forest management plans for potentially certifiable community forestry projects in Esmeraldas province of Ecuador. In Guatemala, Belize, and Costa Rica, CATIE's Proyecto de Bosques Naturales is assisting community-based forestry projects in conducting all aspects of sustainable forest management and is now incorporating a target of "certifiable forest management" as part of the process.

Helping groups and communities to achieve certifiable forest management, and to get certified products into the marketplace, are the objectives of groups such as the Good Wood Alliance (GWA), Forest Partnership (Vermont), Tropical Forest Management Trust (Florida), and Forest Management Foundation (UK). WARP produces the "Good Wood List" and has a field project fund supporting improved forest management and improved wood-processing techniques in countries like Honduras, Mexico, Costa Rica, and Brazil. Forest Partnership is moving to become a nonprofit broker of certified forest products, focusing on the United States. The Forest Management Trust is sponsoring workshops and technical assistance to support potentially certifiable forestry operations in Central America. The U.K.-based Forest Management Foundation, supported by donations from private companies like B & Q (a large DIY or do-it-yourself store), is providing assistance to cover the costs of certification for community-based forestry projects throughout the tropics, with an emphasis on providing products to the European market. Nonprofit organizations with certification programs have also served as a channel for for-profit companies like Ecotimber, Smith & Hawken, or B & Q to provide financial support to underwrite the certification costs for community-based forestry operations in Mexico, Honduras, Papua New Guinea, and Solomon Islands.

With regulatory mechanisms no longer recognized as the most efficient way to achieve positive environmental change, and major regulatory "backlash" in the United States and elsewhere, partnerships are clearly on the rise and show promise for supporting certifiable forest management. However, if NGOs do not have a clear understanding of how their involvement will be publicly presented, or lack sufficient technical skills to properly evaluate each opportunity, their involvement may actually undercut certification by deflecting pressure on a particular company. Such efforts require careful negotiation (usually written contracts) as to how their name will be used and what commitments are necessary from both sides. If done right, the for-profit sector can benefit through the technical expertise, credibility, and relatively low-cost input that NGOs can provide, and the NGOs can gain valuable support for potentially certifiable and sustainable forest management.

Role #5—Lobbying Markets for Certified Products

NGOs have worked hard to push companies and local governments to move their current purchasing away from nonsustainable sources and toward certified operations. Rainforest Action Network (RAN), Friends of the Earth (FOE), Rettet den Regenwald, and Greenpeace have all conducted very visible negative campaigns at stores where unsustainably produced wood products are being sold. Often, their efforts have encouraged bans and boycotts, in some cases encouraging the development of local, regional, or national laws or administrative decrees that mandate the use of certified forest products.

Recently, a number of NGOs, including RAN (Rainforest Action Network 1995) and the Good Wood Alliance (Landis 1993), have begun to focus on pushing buyers to purchase certified forest products. The most visible example of nonprofits stimulating supply has been the 1995 Group effort managed by World Wide Fund for Nature in the United Kingdom. Similar buying groups are underway in Belgium, the Netherlands, the United States, and other countries.

A number of groups have produced guides to help orient consumers about sustainable wood purchasing. In addition to the pioneering Friends of the Earth Good Wood Guide in 1988, the Rainforest Action Network produced *The Wood Users Guide* in 1991 to help wood users identify timber alternatives that do less environmental and cultural damage (Wellner and Dickey 1991) and *Cut Waste Not Trees* in 1995 (RAN 1995). In 1993, GWA and the Rhode Island School of Design joined efforts to organize a "Conservation by Design" traveling exhibition of furniture and other artisanry to illustrate responsible wood use, incorporating the use of certified woods (see *Conservation by Design* edited by Landis (1993), published by GWA and the Rhode Island School of Design to present the exhibits and discuss sustainable forest management and wood use). GWA now produces the "Good Wood List" as part of its quarterly newsletter, *Understory*.

Some NGOs have worked to develop "specification" language that emphasizes preferential use of certified forest products by governments, builders, and architects. Specification language is commonly used by architects, builders, designers, and government agencies when they are buying materials for building projects, through either public or private bidding or procurement practices. Rainforest Alliance, Rainforest Relief, RAN, the Tropical Forest

Foundation, and FOE have either developed and distributed certified forest products specification language or given input to authorities on such draft language, which is then used in public bidding processes for construction projects such as boardwalks, park benches, or other projects in many cities and towns in the United States, Canada, and Western Europe.

In other cases, NGOs have contributed to documents by business associations that give architects or builders advice on environmental policy. An example of this is the American Institute of Architects Environmental Guide Series. These initiatives have tended to support the use of language such as "sustainable forest products," "independently certified," or "FSC accredited" in building supply procurement processes.

Role #6—NGOs as Certifiers

NGOs have played a role as certifiers since the beginning of independent or third-party forest management certification. As noted above, the Smart Wood Program of the Rainforest Alliance was established in 1989 as the first independent forest management certification effort in the world, with its first certification taking place in 1990 (Perum Perhutani in Indonesia as a well-managed forestry operation).

Though there are now numerous incipient certification efforts by local NGOs in countries such as Indonesia, Brazil, Costa Rica, and Mexico (see Box 8.1), the programs of two nonprofits (Rainforest Alliance and Soil Association) and two for-profit companies (Scientific Certification Systems and SGS-Forestry) are the most visible certifiers worldwide at this time. Many

---

Box 8.1

Examples of NGOs directly involved as certifiers

- Consejo Civil Mexicano para Silvicultura Sostenible in Mexico
- Ecoforestry Institute in Oregon
- Ecoforestry Institute Society in British Columbia
- Institute for Sustainable Forestry in California
- Instituto do Manejo e Certificacao Florestal e Agricola (IMAFLORA) in Brazil
- Rainforest Alliance's Smart Wood Program in New York
- Rogue Institute for Ecology and Economy in Oregon
- Silva Forest Foundation in British Columbia
- SKAL in the Netherlands
- Soil Association's Responsible Forestry Program of the United Kingdom

> **Box 8.2**
> **Indonesia's Lembaga Eco-labeling Initiative (LEI)**
> Through a request from the Indonesian Forest Minister, and independent initiative within the NGO community in Indonesia, the Indonesian Eco-labeling Foundation is being established as an independent coordinating structure for voluntary environmental certification in Indonesia, with forest management certification as its first task. An Eco-labeling Working Group was formed in 1993 to establish the LEI; it is composed of representatives from local NGOs, with forestry specialists from government, industry, and universities acting as advisors. LEI has received funding from the World Bank for institutional development, and will develop a system for covering recurrent certification costs through a combination of private industry and government funding. LEI has developed and field tested draft certification standards for natural forest management in Indonesia and will be doing so for tree plantations also. LEI involvement will, at a minimum, focus on standards development, but may also expand to inspector training and accreditation, supervision of assessments and audits, and approval of certifications. Creation of LEI has taken immense amounts of time, as well as negotiations with the commercial forestry sector. (FSC 1995)

NGOs are now considering involvement as certifiers (see Box 8.2 for an example). To make such a decision, there are at least five major issues to consider: (1) scope and depth of activities that will need to be conducted, (2) commercial realities, (3) implications for involvement in other forestry activities, (4) potential conflict of interest, and (5) requirements for certifier accreditation by the FSC.

Scope and Depth of Certification Activities

To implement a certification program, a certifier must develop or have access to the following:

- General and/or region-specific certification standards
- Forest assessment and audit methodologies
- Chain-of-custody assessment and audit methodologies
- Trademark registration and labeling/licensing contracts
- Liability protection systems
- Fee schedules to ensure financial viability
- Accreditation applications to accrediting bodies
- An aggressive public information/promotional effort

Different types of pressures will be placed on an NGO as it begins to work as a certifier. For example, multicountry and/or multiregion auditing is a core requirement for a credible certification program. A small regional NGO in California would face difficulties conducting chain-of-custody inspections in the United Kingdom, for example. International market recognition and coverage, and rapid response, are often critical factors for forest products companies when choosing a certifier. Thus, a company will typically ask a certifier how it is promoting its program and how it will conduct chain-of-custody audits in other countries. Many for-profits are skeptical of non-profits' management capabilities. At least one perspective is that certification is a service provided to the public and forest industry. Providing quick or quality service is not typical of NGOs.

NGOs are also usually asked to make strong commitments to the poor and/or disadvantaged in their work. The FSC requires that even for-profit certifiers seek innovative ways of supporting certification of community forestry projects. However, nonprofits have carried most of the burden in this regard and will likely continue to do so in the near future. This will create pressures not only from a fund-raising point of view, but also because such certifications have historically proven complex, requiring large amounts of planning, assessment, and post-assessment effort.

Commercial Realities

The major structural requirement of a certification program is that, to be successful, it must be run in a businesslike fashion, with quick response capability, a high degree of quality control, strong technical supervision and promotional capabilities, and efficient financial management systems. There is a tremendous amount of administration in a certification program, including the granting and oversight related to the use of certificates or trademarks, the development of promotional materials, the billing of clients, and, most important of all, the scheduling and processing of annual and random certification assessments and audits. The need for business management and marketing skills cannot be overemphasized.

Implications for Involvement in Other Forestry Activities

Because of potential conflict-of-interest issues, certifiers usually must forgo involvement in many historically typical NGO activities. For example, it would be impossible for an organization to be truly objective about an operation for which it has provided services such as forest management planning, biodiversity planning or management, environmental education, or brokering. This means that organizations involved in actual resource management (like The Nature Conservancy) would probably face real challenges if involved directly as a certifier.

There may be ways of working around some of these problems. For example, an NGO might provide general training for "certifiable" or sustainable forest management planning, but not actually provide direct services to a company. However, even in these cases, any certifier will have to carefully guard itself for objectivity.

FSC Requirements for Certifier Accreditation

The FSC has developed an accreditation manual that is being used to evaluate certifiers (Forest Stewardship Council 1995). The process has been based partly on FSC experience with certifiers so far, as well as experience with ISO and organic food certification. FSC requirements are quite detailed and stringent; so NGOs should not assume that achieving accreditation will be an easy task. At this time, accreditation costs to certifiers are not high because of support from donor institutions. However, in the future it is expected that accreditation fees will have to cover all costs for both annual and periodic reviews of certifiers.

Chemicals and Lessons from Organic Food Certification

SKAL (Netherlands) and Soil Association (U.K.) are two nonprofit certifiers that entered forest management certification after working in organic food certification. Organic food certification has gradually evolved to provide flexible, incremental certification, allowing for gradual improvement by companies (decreasing use of chemicals over time). This evolution has been regarded as critical to the future of organic food certification.

The use of chemicals is pervasive in many plantation forestry operations, or even-aged management schemes. For these reasons, the certification of forestry operations that use chemicals presents a challenge to such groups. However, they have tried to respond by developing a more flexible approach in forestry—one that requires minimizing or decreasing the use of chemicals. Nonetheless, some certifiers propose an outright ban on the use of chemicals. The ultimate commercial viability of such a certification approach is unknown.

Structural Options for NGO Involvement

There are basically two different ways for NGOs to become involved as certifiers: working independently, or joining with other NGO certifiers to work as a consortium. One alternative for the latter is the Rainforest Alliance's "Smart Wood Network" (Rainforest Alliance 1995). Rainforest Alliance began this process in 1994 for the following reasons: (1) Region-specific NGO certifiers were at a fundamental competitive disadvantage due to weaknesses in brand recognition and the ability to conduct chain-of-custody auditing worldwide; (2) the Alliance has a mission of support and coordination with local organizations, rather than competition; and (3) its staff felt that NGO presence as certifiers (rather than just observers or pressure groups affecting certification) was crucial for the future of certification as a credible agent for change in commercial forestry, and that without such a coalition, NGO certifiers would be unsuccessful due to commercial forces.

The Smart Wood Network currently includes region-specific affiliate programs in five regions of the United States and Canada, plus country-specific affiliates at nonprofit institutions in Bolivia, Honduras, Mexico, Brazil, and Costa Rica. New affiliates are being explored in Guatemala, the U.S. and Canada, Ecuador, the Netherlands, U.K., Indonesia, and Malaysia. These organizations collaborate in management of the Smart Wood Program, with Rainforest Alliance's Smart Wood staff serving as the headquarters managers and ensuring that all procedures meet FSC's strict requirements.

Summary of Pros and Cons for NGOs as Certifiers

Ultimately, being a certifier requires, and provides, better understanding of on-the-ground commercial forestry. It also requires and provides NGOs with a better understanding of market dynamics, and how to influence commercial interests. With the exception of SKAL and Soil Association, most of the NGOs directly involved as certifiers are also active in other forest activities, including forest policy, education, and research. In fact, most NGOs involved in certification do so because they wish to explore the use of certification as "one tool in the forest conservation and management toolbox." They do not see it as a panacea, merely as a complementary tool to other activities with a strong market focus.

Being a certifier will affect the day-to-day realities of managing an NGO. If an organization has a history of conducting negative campaigns about industry, or acting as a strong advocate in public meetings with government or industry, being a certifier does require moderation of political stances. Thus, being a certifier can impair the advocacy role that NGOs might wish to take. In addition, some people will portray a certifier, even an NGO certifier, as being too involved with industry. Working as a certifier can also mean complicated legal and liability exposure, especially for those operating in the United States.

Despite these difficulties, the presence of nonprofit certifiers should help to ensure that certification has the best chance of being a true agent for change in commercial forestry.

References

Forest Stewardship Council. 1995. *FSC Accreditation Manual.* Oaxaca, Mexico.

Friends of the Earth. 1990. *The Good Wood Guide,* A Friends of the Earth Handbook. London, England.

Landis, Scott, ed. 1993. *Conservation by Design.* October 1993. Easthampton, Massachusetts.

Rainforest Action Network. 1995. *Cut Waste Not Trees—How to Use Less Wood, Cut Pollution, and Create Jobs.* San Francisco, California.

Rainforest Alliance. 1995. "The Smart Wood Network: A Worldwide System of Independent, Non-Profit Organizations for Forest Assessments, Evaluations, Monitoring and Certification." New York, June 1995.

Wallinger, Scott. 1995. "A Commitment to the Future, AFPA's Sustainable Forestry Initiative." *Journal of Forestry* January 1995.

Wellner, Pamela, and Eugene Dickey. 1991. *The Wood Users Guide.* Rainforest Action Network. San Francisco, California.

Chapter 9

Certification as a Catalyst for Change in Tropical Forest Management

Virgílio M. Viana

Although conversion to pasture and agriculture are the most important factors contributing to tropical deforestation (Vincent 1992), the timber industry in general and the international tropical timber trade in particular have also become a target for environmental groups, grassroots movements, multilateral organizations, governments, and researchers concerned with deforestation and forest degradation (WWF 1994). This is because environmentally conscious consumers are quite sensitive to the possible negative effects of logging (Ozanne and Smith 1993) and that, in some regions, logging is a major factor causing forest degradation (Panayotou and Ashton 1992) while in other regions logging fuels forest conversion to pasture and agriculture (Uhl et al. 1991; Veríssimo et al. 1995).

Certification gives consumers a mechanism to avoid products supposedly not produced in an environmentally and socially sound manner, while not eliminating a whole class of products such as tropical timber (Orsdol and Kiekens 1992). Certification can provide economic incentives to good management of tropical forests by increasing and/or maintaining market share and providing premium prices (Crossley et al. 1994). The rationale for certification is that tropical forest management needs to become more economically attractive than other land-use systems to survive (Johnson and Cabarle

111

1993). Unless natural forest management becomes more lucrative than agriculture, pastures, and other competing land uses, tropical forests are likely to be limited to protected areas (Graaf 1986). Tropical forest management has also been limited by social, political, and institutional factors such as land tenure rights and government-sponsored agricultural incentives (Schmink 1989; Schneider 1993). A question that needs to be assessed is the extent to which certification of forest products can provide the necessary incentives to overcome the factors that have limited good forest management practices in tropical forests.

This chapter analyzes the extent to which certification might promote good tropical forest management practices. Here I argue that certification may have a small but important impact on tropical forest management: It can play the role of a catalyst for change. I focus on timber products from natural forests for the sake of brevity. However, parts of the discussion also apply to nontimber forest products and plantation forestry.

The Timber Trade—Tropical Forest Conversion and Degradation

The extent to which the international tropical timber trade affects tropical deforestation is a matter of considerable debate. Varangis et al. (1993) and Thompson (1994) examined the statistics of FAO and concluded that international trade in tropical timber cannot be held primarily responsible for the disappearance of tropical forests since exports account for only a small fraction of the timber produced. Most wood harvested is used for domestic consumption. Total exports of industrial roundwood, sawnwood, and wood-based panels represent only about 14 percent of the total production from developing countries, excluding China. (See Table 9.1.) However, some tropical countries export a much larger part of total production, especially Papua New Guinea (83 percent), Gabon (78 percent), Malaysia (75 percent), Liberia (64 percent), Congo (62 percent), and Indonesia (60 percent) (FAO Yearbook, various years).

The analysis of international trade statistics provides an incomplete picture of the role of the timber industry in the dynamics of deforestation in individual countries. Tropical deforestation is a result of a complex set of factors. For example, worldwide, the leading cause of tropical deforestation is the clearing of forests for agriculture or pasture (Vincent 1992). The decision of forest dwellers, settlers, and ranchers to convert forests into other land uses is a logical and rational decision, given the local socioeconomic conditions,

Table 9.1

Production exports, imports, and domestic consumption of wood products (in 1,000 cubic meters) in developing countries in 1990, excluding China

Product	Production	Exports (%)	Imports	Consumption
Industrial roundwood	300,196	33,860 (11.3)	14,881	281,217
Sawnwood	89,603	10,758 (12.0)	11,163	90,008
Wood-based panels	22,176	13,078 (59.0)	4,593	13,691
Total	411,975	57.696 (14.0)	30,637	384,916

Source: FAO various years

government policies, historical land-use patterns, and economic pressures (Schmink 1989; Schneider 1993).

Linkages between logging and forest conversion have often been overlooked in the estimates of the role of the tropical timber industry on forest conversion. In the Brazilian Amazon, for example, forest degradation and conversion to agriculture or pastures are closely linked to export logging. Uhl et al. (1991) and Veríssimo et al. (1995) have found that logging provided financing for cattle ranchers to clear more forests for new pastures. Logging of mahogany (*Swietenia macrophylla* King) and other valuable timber species also paid for roads that allow settlers and gold miners to encroach farther into the forest. In Southern Pará, one of the main logging roads, the "Estrada Morada do Sol" has been pushed steadily into new forest lands at a rate of 50 to 100 km/yr since 1985, at a cost of U.S.$4,000/km, largely financed by logging companies (Veríssimo et al. 1995). Mahogany logging has also been linked to invasions of Indian reserves by settlers, ranchers, and gold miners in the Brazilian Amazon, because of increased access resulting from road opening (CEDI 1993).

Logged-over forests with no subsequent management are usually biologically impoverished due to extraction and hunting pressures (Nepstad et al. 1992). This is particularly true in regions where forest fires are common. Due to accumulated dead biomass, dry grasses along logging roads, and proximity to pasture and agricultural areas, logged-over forests are more vulnerable to fire (Uhl and Buschbacher 1985). Since forest fires are a major cause of forest degradation in regions such as the Amazon, logging may thus play an important role in forest degradation. Well-managed forests, on the contrary, are less vulnerable to fire and biologically richer than poorly managed, logged-over forests.

The Market Potential for Certified Forest Products

Market size is a key element in determining the potential of certification for influencing current tropical forest management practices. Recent estimates of the size of the market for certified forest products have been made for countries such as the United States. Strategic decisions by large retailers in Europe and North America to purchase certified timber have also provided important indicators of market trends.

Winterhalter and Cassens (1993) conducted a survey in the United States between March and June 1993 on consumer perceptions and "willingness to pay" for "sustainably produced" hardwood products. They used a national sampling service, R. L. Polk and Co., to randomly select 1200 affluent consumers (annual income of U.S.$50,000 or higher). The results showed that 67 percent of this consumer group have changed their purchasing behaviors by boycotting or avoiding products on the basis of environmental concerns (Winterhalter and Cassens 1993). Similar results were obtained by Gerstman and Meyers (1991), who reported that most consumers (84 percent) were more concerned about the environment than they were one year before the study. They also reported that 75 percent of those surveyed would pay up to 5 percent more for environmentally friendly products. Ottoman (1992) estimated that 13.4 percent of consumer products introduced in 1991 were "environmentally positioned," up from 4.5 percent of the total market in 1989. The study of Winterhalter and Cassens (1993) shows that a clear majority (79 percent) place more trust in a label or stamp than on advertisements, brochures and catalogs, or salespeople. In the latter study, most (93 percent) also reported that they would prefer that their furniture originated in a sustainably managed forest. Most consumers prefer temperate wood from North American species; mahogany was the third most preferred species (10.1 percent), after oak (45.9 percent) and cherry (18.3 percent). The majority (68 percent) would be willing to pay more for furniture made from wood that originated from sustainably managed forests. One-fourth (26 percent) would be willing to pay 1 to 5 percent more for assurances of sustainability; one-third (33 percent) would pay 6 to 10 percent more, and nearly one-tenth (8 percent) would pay 11 to 15 percent more (Winterhalter and Cassens 1993).

The conclusions of these surveys indicate that consumers seem to be changing their purchasing behaviors and that environmental labeling is important in their purchase decisions. The studies also suggest a consistent trend in consumer awareness and purchasing behavior that is likely to result in increasing market share for certified forest products. It should be noted that if consumer awareness continues to increase and to be reflected in consumer behavior, the market share and premium prices for certified products are likely to increase in the future.

A few large retail stores in North America and Europe have indicated that they perceive an increasing market for certified timber. For example, in 1991, B & Q, the largest retailer in the United Kingdom that sells temperate (90 percent) as well as tropical (10 percent) timber, announced that from December 1995 onward it would not buy timber products that cannot be "objectively and independently" determined to come from "well-managed sources" (Knight 1992). This decision was based on a number of strategic studies, including a 1989 survey indicating that 76 percent of their customers expressed concern about tropical deforestation and would rather buy certified timber. This is a sizable market: B & Q sales of wood products are worth around U.S.$300 million/year, from wood originating in 34 countries and provided through 115 suppliers (Knight 1992). Another similar case is that of The Home Depot, the United States' largest home improvement center (gross revenues of U.S.$8 billion/year), which has recently made a similar commitment (Eisen 1994). A relevant aspect of the B & Q and The Home Depot cases is that they incorporate temperate and tropical forest products under the same policy. Environmental education campaigns should make consumers aware that buying temperate and boreal timber is not better than buying tropical timber: All forest products, independent of their ecosystem of origin, need to be certified for their sustainability.

A well-publicized project that received the inaugural Earth Summit Award Toward Global Sustainability was Wal-Mart's "Ecostore," inaugurated in 1993 in the United States, designed to be generally environmentally friendly. Wal-Mart paid roughly 25 percent above market prices for 800,000 board feet of "sustainably produced" temperate lumber used to build its 121,264-square-foot store (Clendenning 1993). The premium price could possibly have been lower but for the cost of having to place one-time orders, custom-made to Wal-Mart's criteria. Replication of ecostores by Wal-Mart and other businesses will obviously depend on the return on their investments, directly (sales) or indirectly (publicity). This will be a function of the willingness of consumers to pay higher prices or their preference to shop in such stores.

The cases of Wal-Mart, B & Q, and The Home Depot help counter concerns that certification would be confined to the "boutique end" of the market for small and specialized businesses such as wood crafts and "beauty shops." The boutique market still represents an important player in the future of certification, not because of its absolute market share, but rather because of its leadership role and demonstrated willingness to pay premium prices for forest products.

Most (85 to 86 percent) tropical timber produced is consumed domestically (FAO various years). Although no specific surveys are available, it can be supposed that most of this market is not likely to be willing to pay higher prices for certified timber. However, large countries that domestically consume most of their production of tropical wood products, such as Brazil and Mexico, may develop new markets for certified forest products as their economies improve and environmental awareness increases. There is a need for a quantitative study on the potential of the domestic market of certified products in producing countries such as Brazil.

Most (54.3 percent) of the exported tropical timber goes to Asian importers, primarily Japan (28.1 percent), China (9.2 percent), South Korea (6.2 percent), and Singapore (6.2 percent) (FAO various years). The United States and Europe play a secondary role (7.5 percent and 20.1 percent respectively) (FAO various years). However, because of their consumers' supposedly higher willingness to pay premium prices for certified forest products, the United States and Europe are likely to play a major role in the international market of certified products. Asian markets appear to be generally less sensitive to environmental issues than Europe and North America (Baharuddin and Simula 1994). If indeed the market for certified forest products in Asia is and will remain negligible in the future, this would reduce the potential worldwide impact of certification. However, a quantitative study on the trends of green consumerism in key Asian markets, such as Japan, is needed to make a conclusive analysis.

Certification as a Catalyst of Change

Certification may act as a catalyst of change in several ways. Two of these, providing financial incentives and providing examples of model forestry, are described below.

Financial Incentives

Given information on market size and the premium that consumers are willing to pay, it is possible to evaluate the likely magnitude of the incremental revenues resulting from certification of tropical timber. Gerstman and Meyers (1991) report that 75 percent of consumers would pay up to 5 percent more for certified forest products. Winterhalter and Cassens (1993) give much higher levels of willingness to pay, but for a smaller pool of affluent consumers in the United States. Crossley et al. (1994), based on discussions with NGOs, wholesalers, retailers, and market exporters involved in certification issues, provide a much lower estimate. They suggest that about 10 to 20 percent of the European tropical timber market will be affected by certification, while for the United States the estimate is 5 to 10 percent. The estimated premium price for certified tropical timber is about 10 percent. Table 9.2 presents two scenarios of the size of the "niche" market affected by certification. An optimistic scenario (A) indicates a total value of U.S.$141 million/year of incremental revenues associated with certification, while a more conservative estimate (B) yields a value of U.S.$51 million/year (excluding pulp and paper).

In addition to incremental revenues captured directly from the niche market, estimates for revenues resulting from certification should include recapture of lost markets (U.S.$45 million/year) and averting losses of market share in markets that will require certification (U.S.$400 million/year). The range of all revenues likely to be generated by certification schemes correspond to a 2.5 to 5 percent increase of tropical timber export revenues from developing countries (Crossley et al. 1994). These figures indicate that certification schemes are likely to produce limited financial incentives. However, if consumer behavior continues to change as expected, these figures might grow considerably. The figures presented above provide an approximate value of the financial incentives given to producers and retailers of certified tropical timber by green consumers. It should be noted that not all revenues will revert to forest producers. A share of this sum will go to pay certifiers and to increase the profit margin of retailers (see Chapter 10).

Considering that one of the major constraints to "sustainable" tropical forest management practices are low stumpage fees (Vincent 1992), certification may have an important impact if it results in increases of the stumpage fees paid to forest owners. In the case of Southeastern Amazon (State of Pará),

Table 9.2

Estimates of the increased revenues generated by certification

Location	Market	Scenario A	Scenario B
Europe	3043	114	46
USA	1085	27	5
Total	4128	141	51

Estimates are based on the total value (U.S.$millions) of logs, sawn timber, panels, wood furniture, and wood manufactured imports for the year of 1991 (values based on FAO various years). Scenario "A" assumes a 75 percent and 50 percent market share for Europe and USA, respectively, and a 5 percent premium price paid for certified timber. Scenario "B" assumes a 15 percent and 10 percent market share for Europe and USA, respectively, and a 10 percent premium price paid for certified timber.

for example, the stumpage fee for mahogany is around U.S.$40 per cubic meter, while an export value of about U.S.$700 per cubic meter of sawn boards is reached in Brazil (Veríssimo et al. 1995). Higher stumpage fees may create significant incentives to tropical forest management in some regions, especially those that produce high value timber to European and North American markets.

In addition to the premium prices paid for certified timber (discussed earlier), certification can increase stumpage fees paid to forest owners by shortening the chain of commercialization. The longer the chain, the more difficult and expensive is the system of tracking and labeling products. Companies with short chains of commercialization will have an economic advantage in a market of certified forest products. Reducing the chain of commercialization can generate revenues to offset higher costs of good forest management, labeling, and inspection. The Home Depot, for example, has been able to pay higher prices for certified wood without increasing its price to consumers, simply by reducing the chain of commercialization. Portico, an integrated forestry and industrial enterprise based in Costa Rica and certified by Scientific Certification Systems, has been able to sell its higher value products (such as carved wooden doors) directly to the final retailer (The Home Depot) and thus receive higher prices (Eisen 1994). Cultural Survival has been able to pay higher prices for Brazil nuts produced in extractive reserves in Amazonia partly by eliminating middlemen (Clay 1992).

Model Forestry Initiatives

Considering that the revenues of certification do not appear to be sufficiently great to predict major impacts in the tropical timber market as a whole, what impacts can certification have in tropical forestry management practices? "Sustainable" forest management is practiced in fewer than 1 percent of tropical forests being harvested, but this estimate varies among tropical countries (Poore et al. 1989). In the Brazilian Amazon, for example, the amount of timber produced in well-managed forests is negligible (Viana 1993). A key factor limiting the development of "sustainable" forestry in tropical forests is the scarcity of widely known examples of well-managed forests. Can certification promote the implementation and dissemination of model initiatives of good forest management?

Given information on likely revenues from certification and approximate costs of good forestry management practices, it is possible to estimate the potential size and number of model initiatives of good forest management that could be financed by certification. There are only a few estimates of the costs of good tropical forest management. Baharuddin and Simula (1994) report costs of different estimates in Southeast Asia: U.S.$38 per cubic meter in the Philippines, U.S.$60 per cubic meter in Sarawak, and 20 to 50 percent higher than normal logging costs in a simulation in Indonesia. In the Amazon, IMAZON preliminary reports cite costs of about U.S.$50 to 170 per hectare for Southern Pará (Edson Vidal, personal communication), FUNTAC reports costs 27 percent higher than normal logging costs for the State of Acre (Silva 1992), and Silviconsult (1993) reports costs 30 percent higher than normal logging costs for Guyana.

Considering that a small share (U.S.$10 million) of the total incremental revenues of certification (U.S.$50 to 150 million) goes to forest producers and that the additional costs of implementing good management are about U.S.$200 per hectare (overestimate), this could finance about 50,000 hectares to be brought under good forest management each year. This could mean the availability of about 50 "model forest management initiatives" with an average of 1000 hectares per year each; or a larger number of small-scale operations (e.g., 1,000 operations of 50 hectares per year).

Model forest management initiatives can provide examples of good management practices and become catalysts of change in current tropical forest management practices. Certified examples of good natural forest management practices can have various important roles, with educational, political, and scientific ramifications. Perhaps the most important role is to demonstrate the viability of tropical forest management and its advantages over other competing land uses: pastures, agriculture, and plantation forestry. Cases such as that of IMAZON in Paragominas, Mil Madereiras in Itacoatiara and Tropical Forest Forndation in Portel—all in the Brazilian Amazon—are examples of model initiatives that are serving as catalysts for change in tropical forest management at the regional level. After examples are in place, national governments, nongovernmental organizations, international organizations, and donors may become more effectively engaged in promoting policies to support management of tropical forests. Model initiatives will also provide a focus for applied ecological, socioeconomic, and forestry research. If certification facilitates the development and spread of examples of good tropical forest management initiatives beyond the boundaries of protected areas it will have fulfilled a historic goal in tropical forestry and conservation.

Conclusions

Certification should not be seen as in the interest of only environmentalists of a few industrialized countries; rather it is a mechanism to reward good tropical forest management practices in the interest of forest owners, forest managers and dwellers, as well as governments of tropical countries and concerned citizens of the world.

Certification provides a mechanism for concerned individuals and societies to support "sustainable" forest management. It also represents a new factor in international negotiations over forestry issues. While international negotiations have had limited success in dealing equally well with temperate and tropical forest management, certification is generally seen to include all forest types: natural forests and plantations in tropical, temperate, and boreal regions. This is a very important development since the global implications of poor management practices are not limited to one or another forest type, but rather, to all the world's forests.

The potential financial incentives resulting from certification of tropical timber are relatively small. The expectations about the potential impacts of certification on tropical forestry should not be exaggerated. Certification is no panacea for the complex economic, social, institutional, and political problems that limit good forest management in tropical forests.

Certification may have a small but significant role if it becomes a mechanism to increase stumpage fees. More important, certification may have a historic role in tropical forestry and conservation if it catalyzes the development and spread of examples of initiatives of good tropical forest management in different regions. This may have profound long-term effects on tropical forestry.

Acknowledgment

A version of this paper originally appeared in Ohara, J., M. Endara, T. Wong, C. Hopkins, and P. Maykish, eds. 1994. *Timber Certification: Implications for Tropical Forest Management.* Proceedings from a conference hosted by the student chapter of the International Society of Tropical Foresters, Yale School of Forestry and Environmental Studies, New Haven, Connecticut.

References

Baharuddin, H. G., and M. Simula. 1994. "Certification Schemes for All Timber and Timber Products." Yokohama, Japan: International Tropical Timber Organization (ITTO).

CEDI. 1993. O "ouro verde" da terra dos índios. São Paulo, Brazil: Centro Ecumênico de Documentação e Informação (CEDI).

Clay, J. 1992. Buying in the forests: a new program to market sustainably collected tropical forest products protects forests and forest residents. In K. Redford and C. Padoch, eds. *Conservation of Neotropical Forests: Working from Traditional Resource Use.* Columbia University, New York.

Clendenning, J. 1993. "Promoting Sustainable Forestry Through Certification." B.A. Dissertation, University of Vermont.

Crossley, R., C. A. P. Braga, and P. N. Varangis. 1994. "Is there a commercial case for tropical timber certification?" Paper presented at the Workshop on Eco-labeling and International Trade, UNCTAD, Geneva, June 28–29.

Eisen, M. 1994. "What marketers want from timber certification." Proceedings from Conference on Timber Certification: Implications for Tropical Forest Management. Yale School of Forestry and Environmental Studies, New Haven, CT.

FAO various years. *FAO Yearbook.* Rome: Food and Agriculture Organization.

Gerstman and I. Myers. 1991. *Consumer Solid Waste: awareness, attitude, and behavior study.* Third Annual Nationwide Environmental Survey.

Gillis, M. 1988. "Malaysia: public policies and the tropical forest." Pages 115–164 in R. Repetto and M. Gillis, eds. *Public Policies and the Misuse of Forest Resources.* Cambridge University Press, Cambridge.

Graaf, N.R. 1986. "A silvicultural system for natural regeneration of tropical rainforest in Suriname." Wageningen, the Netherlands, Agricultural University.

Johnson, N., and B. Cabarle. 1993. *Surviving the Cut: Natural Forest Management in the Humid Tropics.* Washington, D.C. World Resource Institute.

Knight, A. 1992. "B & Q's Timber Policy Toward 1995." United Kingdom: B & Q.

Nepstad, D., F. Brown, L. Luz, A. Alexandre, and V. Viana. 1992. Biological impoverishment of Amazonian forests by rubber tappers, loggers, and cattle ranchers. *Advances in Economic Botany* 9: 1–14.

Orsdol, K. G. V., and J. P. Kiekens. 1992. Environmental labeling: a market-based solution for promoting sustainable forestry management in the tropics. In K. Cleaver, ed. "Conservation of West and Central African Rainforests." World Bank, Washington, D.C.

Ottoman, J. A. 1992. Industry's response to green consumerism. *Journal of Business Strategy* 13(July–August): 3–7.

Ozanne, L. K., and P. M. Smith. 1993. Strategies and perspectives of influential environmental organizations toward tropical deforestation. *Forest Products Journal* 43(4): 39–49.

Panayotou, T., and P. Ashton. 1992. *Not by Timber Alone: Economics and Ecology for Sustaining Tropical Forests.* Washington, D.C., Island Press.

Poore, D., P. Burgess, J. Palmer, S. Rietbergen, and T. Synnott. 1989. *No Timber Without Trees.* London, United Kingdom: Earthscan.

Posey, D. 1983. Indigenous knowledge and development: An ideological bridge to the future. *Ciência e Cultura.* 35(7): 877–894.

Schmink, M. 1989. The rationality of tropical forest destruction. In J. C. F. Colon,

F. H. Wadsworth, and S. Brannan, eds. *Management of Forests in Tropical America: Prospects and Technologies*. USDA Forest Service, Rio Piedras, Puerto Rico.

Schneider, R. 1993. *Land abandonment, property rights, and agricultural sustainability in the Amazon*. Washington, D.C. The World Bank.

Silva, Z.A.G.P.P.G. 1992. Análise econômica da exploração sob regime de manejo para a produção florestal sustentada. *Madeira & Cia* November: 6–9.

Silviconsult, S. 1993. "Tropical Forest Management: Position Paper on Certification." Oxford, United Kingdom: SGS Silviconsult Ltd.

Thompson, D. 1994. "Tropical timber: certification and market realities." Proceedings from Conference on Timber Certification: Implications for Tropical Forest Management. Yale School of Forestry and Environmental Studies, New Haven, CT.

Uhl, C., and R. J. Buschbacher. 1985. A disturbing synergism between cattle ranch burning practices and selective tree harvest in Eastern Amazon. *Biotropica* 17(4): 265–268.

Uhl, C., A. Veríssimo, M. M. Mattos, Z. Brandino, and I. C. G. Vieira. 1991. Social, economic, and ecological consequences of selective logging in an Amazon frontier: the case of Tailândia. *Journal of Forest Ecology and Management* 46: 243–273.

Varangis, P. N., C. A. P. Braga, and K. Takeuchi. 1993. "Tropical Timber Trade Policies: What Impact Will Eco-labeling Have?" Washington, D.C. The World Bank.

Veríssimo, A., P. Barreto, R. Tariffa, and C. Uhl. 1995. Extraction of a high-value natural resource from Amazonia: the case of mahogany. *Journal of Forest Ecology and Management.* 72: 39–60.

Viana, V. M. 1993. O selo verde e o manejo de florestas naturais. *Jornal do Engenheiro Florestal.* Associação Paulista de Engenheiros Florestais, São Paulo.

———. 1994. Certification of Forest Products: A Perspective from the South. *Understory* (1): 1–3.

Vincent, J. R. 1992. The Tropical Timber Trade and Sustainable Development. *Science* 256(June): 1651–1655.

Winterhalter, D., and D. L. Cassens. 1993. "United States Hardwood Forests: Consumer Perceptions and Willingness to Pay." Department of Forestry and Natural Resources, Purdue University.

WWF. 1994. "The East Asian Timber Trade." Godalming, Surrey, United Kingdom, World Wide Fund for Nature.

Chapter 10

Economics of Certification

Markku Simula

Certification is a policy instrument that should be effective in contributing to the achievement of its two main objectives: sustainability of forest management and market access (Baharuddin and Simula 1994). As an economic incentive, certification should provide net benefits at the level of a forest management unit, which is the subject of certification. To make the process work in practice, all the various phases of the production and distribution chain should obtain net benefits. Impacts may also be neutral or negative. In the latter case, the issue of compensation arises.

Experience in certification of forest products is still limited and the procedures are evolving. The situation does not yet lend itself to a rigorous economic analysis. This chapter is therefore exploratory by nature, focusing on the identification of relevant issues and reporting on available information on possible benefits and costs involved.

Economic Issues Related to Certification Criteria and Indicators

The criteria and indicators used in the assessment of forest management may be expressed in terms of standards. They largely determine the cost implications of certification. From the economic point of view, it is important that the indicators are goal-oriented and lead to efficiency in implementation. Standards set on an international level concerning different types of forests are, by definition, general by nature. Each country has its own characteristics, which should be considered when the respective national-level criteria and indicators are established.

The incremental costs of "good" forest management depend on the difference between the standards to be applied in assessment and the current status of forest management.

From the economic point of view, it is important to distinguish among the certification standards and those defined in the legislation, rules, and regulations of the country in question. Incremental costs can be considered due to certification, while the government standards should be met in any case by forest managers.

Certification is a voluntary activity, and therefore it can be assumed that the standards of assessment are set above those defined by the government. However, sustainability in the broad sense of economic and social benefits and maintenance of ecological functions has been given political recognition by governments (through documents such as UNCED Forest Principles, ITTO Guidelines of Sustainable Forest Management, the Resolution of the Helsinki European Ministerial Conference on Forests, and others). It is, therefore, foreseen that in the long run the government standards and certification criteria may tend to converge, and the issue of incremental costs due to certification may gradually lose its importance. Nevertheless, in the short and medium term the question of incremental costs is relevant (see Figure 10.1).

Let us take an example in biodiversity conservation. From the economic point of view, the implications of ensuring the continuous existence of a particular species are quite different from achieving a certain abundance of the same species. In a simplistic example, if a key habitat is needed for protecting a threatened species, in management decisions we have to ask whether all of these habitats should be set aside, or whether only a sufficient amount of them should be protected to ensure sustainable population levels.

If information on biodiversity is adequate, there is less need to apply the precautionary principle in management. More accurate criteria can be applied that are likely to represent lower input requirements from the economic point of view. This is a justification for investment in research on biodiversity.

In the same way as sustained-yield timber management is not always a practical concept in smaller management units, some biodiversity conservation objectives may not be realistic at the small scale. For instance, in boreal

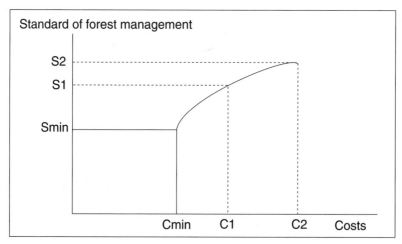

Figure 10.1
Costs of Certification. It can be assumed that marginal benefits decline exponentially in relation to marginal costs after a minimum limit of an environmental or social standard (threshold value) has been met (Smin in Figure 10.1). This means that raising the level of criteria (e.g., from S1 to S2) may require a proportionally higher increase in costs (from C1 to C2), assuming the threshold has already been exceeded.

forests the vegetation mosaic shifts continuously over time and therefore biodiversity planning and monitoring should be applied over large enough units to be meaningful.

If certification criteria and indicators are set on the level of biogeographic zones or other relevant regional units (within countries), they are likely to be economically more efficient than national-level or broader international standards. This is because such indicators can be based on specific local conditions, and they can be related to specific objectives in biodiversity conservation or social development.

This view derives from the forest management objective of certification, but it is clear that any location-specific criteria have to address the general sets of forest management standards, which provide a basis for international harmonization. In practice, a balance has to be found among the following requirements of the criteria: (1) compatibility with the agreed international principles and criteria, (2) effectiveness and efficiency in local (national) conditions, and (3) applicability on the level of forest management unit.

Another issue related to criteria is whether they should be prescriptive. There appears to be strong preference for a flexible nonprescriptive approach to promote the continuous improvement of forest management operations without specifying a constant target level of achievement (for example, FSC Principles and Criteria). This characteristic is also typical in the existing eco-labeling schemes. From the economic point of view, this would mean that incremental costs are likely to change over time depending on the cost implications of changes in assessment criteria.

Costs Related to Certification of Forest Products

Both indirect and direct costs are involved in certification of forest products. (See Table 10.1.) Indirect costs refer to activities necessary to (1) achieve the assessment criteria of forest management and (2) establish adequate management and information systems so that a reliable verification can be carried out at reasonable costs. We can also call the latter component information costs of certifiability. Direct costs are needed to cover the actual certification operation; they are paid to the certifying organization.

Setting up certification schemes generally requires significant investments in the establishment of the institutional framework (accreditation and certification), definition of criteria and indicators, human resource development, development of administrative procedures, and so on. These costs can vary extensively depending on local conditions. The existing certification systems are self-financing; therefore, their investment and running costs are expected to be covered by the fees levied on their services. The investment costs of setting up certification schemes are not discussed here due to lack of relevant information.

The information costs of certifiability and the direct costs of certification are linked. In general, the better the documentation and internal inspection systems, the lower the costs to be paid for external inspection.

Incremental Costs of Forest Management

The incremental costs (including foregone benefits) of forest management may derive from five different sources: (1) set-aside areas, (2) lower yield per areal unit in harvesting areas, (3) additional silviculture and harvesting costs, (4) additional costs of planning and monitoring, and (5) different distribution of costs and benefits over time. An accurate assessment of the above costs would require site-specific analyses in relation to the current government regulations and the certification criteria to be applied in those conditions. The difference between the two could then be attributed to the incremental costs of forest management due to certification. As such an analysis is not possible at present, the following discussion is based on the compari-

Table 10.1
Costs related to certification

A. Indirect costs
  1. Incremental costs of forest management to meet certification criteria
    (i) Investment costs
    (ii) Silviculture
    (iii) Harvesting
    (iv) Other management costs
      • conservation areas
  2. Information costs of certification
    (i) Forest management
      • resources inventories and surveys (timber, biodiversity, soil, waste, and so on)
      • socioeconomic surveys
      • forest management planning
      • recording and reporting on activities carried out, production volumes, and so on
      • internal inspections and other management costs
    (ii) Chain of custody
      • marking of logs and products
      • recording and reporting
      • additional costs of transportation, storing, processing, and distribution
      • internal inspection and other management costs
B. Direct costs
  1. Application
  2. Inspection (initial)
  3. Annual auditing
  4. Fixed fees (royalties and other)

---

son of the existing situation and the perceived criteria of good forest management that would also qualify for certification.

The needs for set-aside areas for key biotopes or landscape features depend mostly on the local conditions. General rules for the extent of areas to be allocated for nonproductive uses are subject to debate and the range appears to vary from 5 percent up to 50 percent. Among the existing certifiers, Rainforest Alliance (1993) proposes that roughly 10 percent of the total area under forest management (excluding stream and roadside buffers) be designated as "conservation zones."

The voluntary action in forests owned by forest industry companies in the Nordic countries (large estates where landscape factors can be considered) has been in the range of from 5 to 15 percent in terms of area. The impact on wood production has been less, as the productivity of the set-aside areas has been lower than average. It can be expected that the share may not be less in temperate or tropical forests due to their more extensive biodiversity: In Brazil the forest law requires that 20 percent of the plantation areas be set aside as "legal reserves" that are not used for harvesting of wood.

Lower harvesting yield per unit area in natural tropical forests may be due to "low impact logging" (coupled with harvesting of fewer trees) to reduce damage on the remaining vegetation. In temperate and boreal forests, lower yield may be due to leaving more trees behind in regeneration cutting areas for biodiversity conservation. The short-term economic sacrifice could be somewhat compensated in the tropical forests by higher overall yield of timber in the long run and reduced damage to nontimber values (Bach and Gram 1993). These arguments can also be considered relevant in temperate and boreal forests in many conditions.

Silviculture and harvesting costs, in some cases, appear to increase when nontimber values are considered in forest management, but savings can be achieved. In reforestation, use of natural regeneration can be often applied to cut down planting costs, while cleaning and tending costs may be reduced if a mixed species composition is targeted. The actual harvesting costs may not, however, be significantly higher, as Carlén (1994) found in Sweden. In the tropics, Hendrison (1989) found that the costs were not increased substantially when controlled harvesting was introduced as part of the CELOS management system in Surinam. Leslie (as cited by Bach and Gram 1993), however, points out that sustainable management should require very low impact harvesting if industrial timber is produced in natural tropical forest. The existing market and commercial conditions do not provide short-term incentives for forest managers, who should be able to compensate for high initial investment and a substantial reduction in the yield.

The additional costs of planning and monitoring can be significant, particularly in the initial stage. The necessary activities typically include mapping, inventory, management planning, road planning, preharvest enumeration, logging planning preparation, skid trail planning, sample plots establishment, post-harvest inventory, environmental impact studies, and so on. In plantation forests the costs are substantially lower.

It is obvious that improved planning reduces costs. In the case of the mixed Dipterocarp Hill Forest in Sarawak, Marn and Jonkers (1981) concluded that considerable savings can be achieved in harvesting costs through better planning, and these savings would be larger than the planning costs. Additional planning costs in the temperate and boreal forests would be mostly due to the need to expand the scope of traditional forest inventory and resource surveys to cover biodiversity and landscape aspects. An initial cost would be involved to establish necessary databases to cover all relevant environmental and socioeconomic characteristics of the area. Biodiversity would also have to be included in a systematic manner in forest management planning. Only marginal additional costs may be involved when the necessary training has first been provided to planners and adequate databases have been established on national, regional, and estate levels.

In their assessment of incremental forest management costs, Bach and Gram (1993) conclude that the different distribution of costs and revenues in sustainable management can be an economic constraint. The question of lower yield discussed above is partly a distribution issue. The problem is aggravated because part of the additional costs during the initial period would increase when the foregone benefits would also occur. Long-time horizons are needed in the economic assessments, which makes evaluation of risks difficult.

It is concluded that incremental costs would be unavoidable in most situations when forest management is improved. The main reasons are foregone benefits. The prevailing levels of forest management, infrastructure, human resources, and information systems vary extensively by countries, and therefore the incremental costs also vary. It is not possible to make estimates on the amount of incremental costs caused by certification based on the available information, as the existing studies are not related to the achievement of specific standards. (See Utsunomiya 1995 for a summary of nine studies on the tropics.)

Costs Involved in Chain-of-Custody Control

Several solutions exist for tracking logs from the forest to the mill yards of processing plants or export ports. (See Chapter 6 in this volume.) Marking and recording individual saw and veneer logs is a common practice in many tropical countries. The chain of custody is usually well controlled in cases where the industry procures its raw material from its own forests or concession areas and processes the whole wood output itself. However, the control of the chain of custody of forest products is a complex issue in many countries, and cost implications can be substantial. The following situations should be considered in the design of appropriate systems:

1. The roundwood markets may be a complex network of sellers, buyers, and wholesalers. Some of them are reselling their unwanted timber assortments. (For example, if the buyer is a sawmill that buys standing timber, pulpwood has to be marketed to other users.)

2. Roundwood markets may be fragmented if they are dominated by a large number of small private owners who typically sell small quantities of wood at a time.

3. Industrial residues (chips and sawdust) are traded between companies.

4. In the case of imported roundwood, the local rules and regulations in the country of origin have to be respected; these may not be compatible with certification criteria applicable in the importing country.

5. The final product (such as a piece of furniture) may contain several wood products of different origins.

The incremental costs of chain-of-custody of logs depend on the adequacy of the existing practices in marking and recording. In simple operations minor modifications to the current practices would be needed. Rather expensive solutions are being promoted for large-scale, complex systems that may require online computer-accessible tracing. The justification of such investment may be found through other benefits than purely certification (such as reduced corruption and tax evasion, and better inventory management).

From the cost perspective it is important to distinguish situations where certification of the chain of custody is based on tracking of individual logs, pieces of wood products, and fiber and control is based on input flows of raw materials, in cases where the company is obliged to buy wood originating from both certified and uncertified forests. For instance, the certification program of Soil Association (1994) requires that at least 90 percent of timber used in a product has to come from an approved source to qualify for certification.

From the processing industry's point of view, a key issue is whether the certified timber has to be kept separate, from the wood yard through processing to the warehouse of sales products. If this is imposed on existing mills, excessive investment costs would probably be needed and operating costs would increase. In other words, parallel production of certified and uncertified products would have to be practiced.

In the wood yard the storage areas would have to be increased, and in sawmills log and lumber sorters should be extended to cope with various dimensions and grades of certified logs. Processing of different types of raw materials could be theoretically controlled, but operational efficiency is likely to suffer when the number of wood assortments is doubled. More space and incremental transport costs would be needed for keeping separate green and dried lumber before and after kilning and in further processing. By-products should also be separated, which would require investment in storage space and conveyors.

In pulp and paper mills, parallel production would also require increased investment and operating costs reducing overall efficiency. Additional costs would occur in the wood yard and intermediate storage of fiber in different phases of the process, depending on whether the process was based on batch or continuous digesters.

Even more important than additional costs is the fact that the existing mills have not been designed for parallel production, and many of them cannot introduce such a practice because of the lack of space in the mill site. Furthermore, the environmental burden of industrial and transport operations might increase.

The situation would be different if the arrangements were designed before new mills are established. The economic service life of main machinery and equipment is typically about 15 years in sawmilling and 20 years in pulp and paper production; therefore, structural changes would be slow.

Based on this background it is concluded that a cost-effective way, with minimum additional environmental burden, to solve the problem of controlling wood and fiber flows in the various processes of primary forest industries would be through the records of the incoming raw materials, based on which appropriate decisions on certification could be made. The whole output may qualify for certification when a certain minimum threshold level has been met, or the system could be based on the actual input-output ratios (for example, 50 percent certified raw materials would allow certification of 50 percent of output). In secondary wood and paper processing, there are somewhat better possibilities to track different types of raw materials throughout the process.

In trading, the additional costs of chain-of-custody control are likely to remain limited when the initial investments in the information systems have been made. The barcode-based inventory management capabilities are becoming more common in wood trade, for example in the United States, and these systems can easily incorporate the data requirements of chain of custody, particularly when sellers and buyers are linked with each other (Vlosky and Ozanne 1995).

Direct Costs of Certification

Reliable data on the direct costs of certification are limited to some certifiers who have been willing to release such information, which, in the case of private companies, is considered confidential commercial data. The current rates may not be representative for the future if certification becomes a major business activity subject to competition between certifiers. Their pricing policies will also be influenced by long-term considerations, particularly if the same company is providing services both in certification and consulting to assist clients in implementing the recommendations of the assessment exercises.

Forest Management

The costs of the initial inspection depend on the information available and what kind of team (number of persons, expatriate or local) is involved. The existing certifiers are located in the United States, the United Kingdom, and the Netherlands. Fielding of assessment teams outside these countries represents a major additional cost element. To eliminate this disadvantage, at least one certifier, Rainforest Alliance (1993), has a policy to use local specialists as much as possible. Certifiers in a number of developing countries are now close to becoming operational.

Another major factor having influence on the unit costs of inspection is the size of holdings. This can be a major obstacle if holdings are small. They could be certified using sampling based on a regional unit, which may be a cooperative, an association, a municipality, a district, or some other unit.

Theoretical unit cost functions are depicted in Figure 10.2. Heaton (1994) confirms the form of the assumed cost function in her estimates:

Estate size (ha)	Unit cost US$/ha
5,000	1.30
100,000	0.24
600,000	0.08
several million	0.01

The following figures on inspection costs are available: U.S.$5,000 to 75,000 per assessment depending on the consultancy time and travel costs (Donovan 1994); U.S.$0 to 3,000 per assessment if subsidized by the program, U.S.$3,000 to 4,500 per full-cost assessment (Heaton 1994); U.S.$31,000 annually for a large overseas operation for a 500,000-hectare operation (SGS Silviconsult 1994).

Soil Association charges 1 percent of sales value in large forest operations, but in smaller operations the annual fee varies from about U.S.$225 upward per estate depending on the sales value.

In addition to initial inspection and annual audits, application fees and annual fixed fees may apply, but they may not represent significant costs (Heaton 1994).

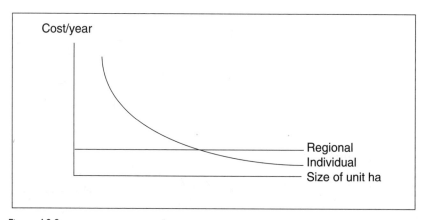

Figure 10.2
Theoretical Unit Cost Functions. It can be assumed that the unit cost of certification for individual forest owners declines as the size of land holding increases, while the unit cost of certification implemented on a regional basis would theoretically remain constant regardless of acreage involved.

Chain of Custody

There appears to be even less definition of the costs of chain-of-custody certification than in the case of assessment of forest management. In sample audits the costs may be low. According to Heaton (1994), they vary from U.S.$100 to 2,000 (average U.S.$500) per annual audit and company audit. Substantially higher costs have also been suggested—for example, 1 percent of the "border price" for imported wood in Switzerland (SGS Silviconsult 1993). If the recording and reporting requirements are integrated in the information systems of the companies, the chain-of-custody audits may not prove to be costly activities. The costs may be comparable to standard auditing costs.

Current information on certification costs may not be representative for the future if certification becomes a major activity. This is due to several factors: (1) There are very few experienced certifiers in the world, and the areas and volumes have been more experimental than routine operations, with few exceptions; (2) many issues still lack definition as many schemes are still in the planning phase; and (3) competition between certifiers has not been a major factor influencing costs.

It is possible that future participation in certification schemes may be priced in the same way as existing eco-labeling programs, which typically have application and annual fees supplemented by a royalty on the value of sales (with a possible ceiling). In addition to these, direct inspection and auditing costs must be paid separately to the certifier. For instance, Soil Association's annual licensing fee is 1 percent of added value or labeled sales in larger operations.

Economic Benefits

The incremental economic benefits of certification of forest products may derive from increased forest yield as a result of improved forest management, higher sales revenue because of larger market share (sales volume) and possibly higher prices, and improved efficiency in forest harvesting, industrial processing, and distribution and marketing.

As pointed out earlier, improved management can, in the long run, be expected to result in higher yield. In the cost-benefit assessment the choice of the discount rate will, in general, have a decisive impact on the result. The marginal impact of certification on yield will also depend on how far the national standards and the certification criteria have been set from each other. Due to lack of definition of general standards, it is not possible to assess the magnitude of these benefits.

Another aspect is to what extent certification will promote forest management improvements that would not happen otherwise. This impact can only be assessed based on observation over time.

The market benefits are discussed in other chapters of this book; only a few remarks are made here. Some market segments in environmentally sensitized markets are certainly willing to pay a "green premium," but it has been cautioned to include such an additional revenue in broader policy assessments (Baharuddin and Simula 1994).

In the case of tropical timber, Crossley et al. (1994) estimate that (under a set of fairly optimistic assumptions) the potential incremental revenues associated with certification could amount to 4.8 percent of the respective export revenue of developing countries. They conclude that this would represent only limited potential incentives for forest management.

Improved cost-efficiency can be expected from the environmental management systems that will be established by forest managers, industry, and trade to qualify for certification. The marginal contribution of certification can again be expected to come from the fact that the improved management systems would be adopted by the sector faster when induced by certification.

A potentially more important source of economic benefits is the possible elimination of trade intermediaries between procedures and final consumers of forest products. There are a number of reported cases of such examples, but it is not yet possible to assess in broader terms the potential economic bene-fits of shortened distribution channels. In countries where illegal practices are common, efficiency gains can also be expected from better control of wood flows through certification.

Conclusions

The economics of certification is an issue for the forestry sector, but it is still too early to assess the effectiveness of this policy instrument in achieving its two main goals: improved forest management and market access.

Certification will undoubtedly increase costs, but in many cases the true in-cremental costs may remain limited. Available information does not enable reliable cost impact assessment due to lack of definition of many parameters, certification of forest products still being in a nascent state. However, the cost implications will vary depending on countries and local conditions.

There is a risk that the main benefits will be found in market access and that only limited incremental impact will be generated in forest management. This should be considered in the design of future certification schemes.

References

Bach, C., and S. Gram. 1993. *The Tropical Timber Triangle—A Production-Related Agreement on Tropical Timber.* KVL. The Royal Veterinary and Agricultural University. The Netherlands.

Baharuddin, H. G., and M. Simula. 1994. *Certification Systems of All Timber and Tim-ber Products.* ITTO. Cartagena de Indias.

Carlén, O. 1994. *Kostnader för naturvårdshänsyn inom privatskogbruket.* Swedish Uni-versity of Agriculture.

Crossley, R., C. A. Primo Braga, and P. N. Varangis. 1994. *Is There a Commercial Case for Timber Certification?* The World Bank.

Donovan, R. Z. 1994. *Strategic Options for Initiating Voluntary and International For-est Management Certification in Bolivia.* Proyecto BOLFOR. Santa Cruz.

Heaton, K. 1994. *Perspectives on Certification from the Smart Wood Certification Pro-gram.* New York.

Hendrison, J. 1989. *Damage-Controlled Logging in Managed Tropical Rain Forests in Suriname.* Wageningen Agricultural University.

Marn, H. M., and W. Jonkers. 1981. "Logging Damage in Tropical High Forests." Working paper no. 5. The Government of Malaysia/UNDP/FAO, Kuching.

Rainforest Alliance. 1993. "Generic Guidelines for Assessing Natural Forest Management."

SGS Silviconsult. 1993. "Swiss Certification Program." Field Report. Oxford.

SGS Silviconsult. 1994. "Tropical Forest Management." Position Paper on Certification. Oxford.

Soil Association Marketing Company. 1994. "Responsible Forestry Standards." Bristol.

Utsunomiya, S. 1995. "Cost Analysis of Sustainable Forest Management in the Tropics." Master's Project. School of Environment. Duke University.

Vlosky, R. P., and L. K. Ozanne. 1995. "Chain of Custody Vital to Certification Process." *World Wood* 122: 2, 35–36.

Chapter 11

Potential Inequalities and Unintended Effects of Certification

Chris Elliott and Virgílio M. Viana

Certification, like any policy instrument, may have "unintended" or "negative" effects. Unintended effects may be either positive or negative, whereas negative effects run counter to the objectives of the policy instrument. This chapter will concentrate on negative effects, which can be classified into several categories. Most negative effects involve barriers to certification. The implication of this is that certification itself does not usually lead to negative effects, but that there may be hidden barriers to getting certified, which prevent some operations from obtaining the benefits that may be associated with certification.

Barriers to Trade

There is strong debate over the potential effects of certification and ecolabeling as barriers to trade. The World Trade Organization (WTO) provides the main intergovernmental forum for discussing international trade issues and establishing rules for international trade; WTO is the forum where the possibility of certification being challenged as a trade barrier will most probably arise.

The rules of the General Agreement on Tariffs and Trade (GATT), the agreement on which the WTO is founded, have been interpreted as preventing the use of trade measures that discriminate among similar products on the basis of the method of production. That is, an importing county cannot use trade measures such as bans, quotas, or tariffs to back up requirements for a particular kind of forest management in exporting countries. Some export-related trade measures such as tariffs may be "GATT legal," so long as they are applied only to timber produced in the exporting countries. In other words, a country can use some trade measures to encourage more environmentally friendly forest management in its own jurisdiction, but not in any other country's jurisdiction.

However, it is normally considered that if eco-labeling or certification is voluntary (an initiative of NGOs or the private sector rather than mandated by government), it should not lead to trade distortions, because in the end it is the individual consumer who decides whether to purchase a particular product, rather than the decision being made for him or her by import barriers. A recent study by Dutch lawyers concluded that timber certification would not be "GATT-illegal" as long as it was voluntary (Droogsma 1991). However, if certification were limited to tropical timber it would be politically unacceptable to many developing countries, as statements from Malaysian and Indonesian officials have indicated (ITTO 1994).

The situation has become more complicated since the completion of the Uruguay Round negotiations, which led to the creation of the World Trade Organization (WTO) to replace the GATT, and the Agreement on Technical Barriers to Trade. A recent study by UNCTAD (United Nations Conference on Trade and Development) indicated: "It may be possible to argue that, even in cases where the eco-labeling schemes are private, the provisions of the GATT/WTO Agreement on Technical Barriers to Trade request countries to at least make 'their best endeavors' to see that nongovernmental bodies adhere to the procedures adopted under conformity assessment systems" (UNCTAD 1995). Conformity assessment systems have been developed by the International Organization for Standardization (ISO). The implications of this for timber certification are still unclear, but it would appear that certifiers would be well advised to set up mechanisms that avoid discrimination between operations on the basis of geography or forest type, and that are compatible with ISO procedures.

Even considering that certification can be "WTO-legal" given its voluntary nature, certification may still be seen as a barrier to trade if it excludes individual producers (or countries) from a significant share of the world market. To minimize such effects, access to information and services of certification should be made available to producers as widely as possible.

Developing Countries

Developing countries are becoming increasingly concerned that although eco-labeling schemes, including timber certification, primarily have environmental objectives, they may have trade impacts. Certification schemes in major timber importing countries could lead to discrimination against foreign producers, particularly those in developing countries, because of the way they operate, and may end up acting as nontariff barriers to trade. This could arise in situations where certification is formally voluntary but is in fact a condition of market access and thus de facto becomes obligatory.

This can happen in several ways: Producers in developing countries may have difficulty getting access to information on the requirements of eco-labeling schemes and may be faced by the requirements of different schemes in different markets.

Further, domestic producers may be more able to influence the setting of criteria and selection of product categories to be labeled than foreign ones. If standards for certification are set in the importing country, they may be unrealistically high for developing countries to meet, or may neglect specific local conditions in which developing countries have a comparative environmental advantage. This could easily occur if standards are set in temperate countries for tropical forest exporters. A case in point has been raised by Brazilian pulp and paper manufacturers who claim that European Union (EU) eco-labeling criteria requiring partial use of recycled paper discriminates against plantations in the tropics that are producing virgin pulp. This problem extends well beyond timber certification, but problems in other forms of labeling will have implications for, and effects on, timber certification: UNCTAD estimates that around 45 percent of the imports in broad product categories that have been identified for eco-labeling in the EU originate in developing countries (UNCTAD 1995). In a recent review of the trade impact of eco-labels (da Motta Veiga et al. 1994), a team of Brazilian specialists concluded that "the environmental regulations in OECD countries—particularly the European Union—are generally perceived by Brazilian exporters as a threat to the maintenance of positions won by Brazilian products in the markets of these countries over the past two decades."

Viana (see Part III, "Geographic Regions") has also indicated that although most certifiers at present are located in developed countries, many certified forests are in developing countries, so there is risk of a trend toward the concentration of, the expertise on, and the financial benefits from, certification in developed countries. Certifiers in developing countries may find it hard to enter the market under these circumstances. There should be mechanisms to support the establishment of certifiers in developing countries.

Donovan (see Chapter 8) provides a promising example of partnership between "northern" and "southern" certifiers through the Smart Wood Network.

It is also important to note that the impetus behind eco-labeling in developed and developing countries may be somewhat different. In developing countries, access to the international market with more and richer "green consumers" is often a key incentive for producers. Access to the international market can result in a significant price premium: The community of El Pan in northern Ecuador was able to export some timber to the green market in the United Kingdom through the Ecological Trading Company in 1992 for a price of U.S.$400 per cubic meter compared to the domestic market price of approximately U.S.$20 per cubic meter in Ecuador (Chris Cox, personal communication). On the other hand, in developed countries the main impetus may be accessing market niches domestically.

An ITTO report (1994) notes that because of generally weaker forest administrations and infrastructure, developing countries will often be in a less favorable situation for certification than developed ones. Increased funds may be required to reach performance levels needed for certification, and these are usually lacking. Chain-of-custody may also be more difficult to establish in some tropical countries, where purchasing is often done at the roadside and numerous intermediaries are involved, often working informally.

Small and Medium Enterprises

The UNCTAD study showed that small and medium enterprises have many of the same problems mentioned above. The situation is amplified if the enterprise is in a developing country, whether it be a for-profit corporation or an indigenous forestry cooperative (UNCTAD 1995).

The costs involved in adapting to the requirements of certification in terms of the use of specific technology and raw materials, as well as the costs of certification and tracing, may be prohibitive.

The ITTO report on certification (1995) has noted:

> As the cost of certification is essentially a fixed cost, large-scale forest owners or concession holders could absorb the additional cost more easily than small forest owners due to economies of scale. This will be particularly true if the smaller owners have less secure tenure as is often the case, and therefore less incentive to invest in the forest resource. Special arrangements (e.g., regional or district approach) might be required for communal lands and small forest owners to avoid penalizing them through additional costs of certification.

It is questionable whether the costs of certification are actually fixed, but in any case it is clear that small-scale operations will tend to be at a financial disadvantage when they absorb the costs.

In addition to scale, the level of integration is another important issue. Large, vertically integrated operations would have an advantage over small, less integrated ones. This might place countries such as Canada, where forest operations are large-scale and companies often vertically integrated, at an advantage compared to countries such as France, where forest holdings tend to be smaller and companies less integrated.

Geographic, Market, and Social Issues

Certification is likely to be more important in markets with higher environmental awareness, particularly Northern Europe and North America. Producers dependent on these markets (for example, from Africa or the Nordic countries) and domestic industry in these countries could be expected to draw the benefits and suffer the costs of certification before those in other regions. If certification "takes off," producers in these regions will have had a head start that will place them at an advantage compared to those exporting to markets where certification is currently not an issue. Again, developing countries that may not have a significant domestic "green market" could initially be placed at a disadvantage compared to developed countries.

A green premium obtained in some markets (although this is probably a limited phenomenon—see Chapter 10) could have negative impacts on lower-income consumers who might be faced with higher wood prices if retailers choose to sell only certified timber. In countries like the United States, where wood is often used in home construction, this could have social impacts. In addition, the elimination of intermediaries mentioned earlier could have impacts on employment.

It is also not clear that market benefits from certification that accrue to retailers in consuming countries will necessarily translate into benefits for the producers of timber. A situation could arise where retailers in developed countries are paid a higher price for certified timber but do not pass on the benefits of this to producers in a tropical country, or where the costs of certification absorb the market benefits, with no net benefits accruing to forest managers.

Indigenous communities involved in forest management may be faced with the additional problem that certification is based on written management plans, which are not used in many indigenous communities that depend more on oral transmission of traditional knowledge. It would be extremely unfortunate if this became a barrier to certification, as many examples of traditional forest stewardship are to be found in indigenous communities. In these cases, certifiers should be encouraged to use culturally appropriate approaches that do not discriminate against societies in which oral traditions predominate.

Substitution

For most end uses timber can be replaced by other products. This is referred to as "substitution" and may occur at two levels: (1) between timber and other products and (2) within timber (e.g., between tropical and temperate timber).

Some timber industry representatives argue that the debate about certification may tend to give consumers the impression that wood is not an "environmentally friendly" product (unless it is certified) and that they should prefer alternatives such as plastics, steel, or aluminum, which are less controversial. They say that substitution for wood by these products is increasing in some markets, and that certification will not be effective in combatting this trend because it focuses on one part of the life cycle of the product (see Chapter 5)—forest management, where these substitutes cannot easily be compared to wood. They argue that certification should be based on a full life-cycle assessment taking into account transportation, processing, and disposal.

Because the timber market is a global one, and because of improvements in processing technology, substitution appears to be common between tropical and temperate timbers for many end uses. On the one hand, this means that producers who make the necessary investments to obtain certification could be subject to competition based on prices, from other producers who have not been certified. This can work both ways however, and certification might be a means for some producers to enter the international market, or niche markets, and obtain higher prices than they could domestically.

Certification schemes will need to be carefully designed to avoid inadvertently favoring tropical or temperate timber, for the reasons mentioned at the beginning of this chapter. For example there is a tendency to assume that boreal forests are easier to manage than tropical ones because of lower species diversity and greater adaptation to disturbance.

If this assumption were accepted, it might lead to relatively lower standards being set for boreal than tropical forests. However such assumptions are questionable in several ways: First, there are many different types of tropical forests, some of which, such as those in floodplain areas, or dry forests that burn more often than previously thought (Sanford 1985), are more adapted to natural disturbance than previously thought. Second, relatively low levels of diversity or endemism in a particular forest type should not reduce the requirements to maintain viable populations of these species.

Certification does not specifically address wood quality, since it is an assessment of the quality of forest management. It is thus conceivable that a piece of certified timber might have lower structural properties than noncertified timber. It is important that this be clear to consumers to avoid confusion.

Natural Forests and Plantations

Just as it is often assumed that temperate and boreal forests are "easier" to manage than tropical ones, it is often considered that plantations are "simpler" to deal with than natural forests, because of the uniformity of species composition and age classes usually associated with plantations. However it can be argued that this also is a generalization. The establishment of plantations has been highly controversial in many countries, particularly when primary forests have been cleared for this purpose, or "common lands" taken over from traditional populations for planting. This has resulted in both social and environmental impacts. In short, the issues of land conversion to establish plantations cannot be separated from assessment of the sustainability of the plantations. In addition, natural forests tend to produce a greater range of goods and services, ranging from nontimber forest products to watershed protection. If certification were to discriminate against natural forest management and in favor of plantations, there is a risk that these benefits could be placed under threat in some circumstances.

Conclusions

Certification could have a number of unintended negative impacts, particularly on small-scale operations managing natural forests in tropical countries. It will be necessary, in some circumstances, to develop specific mechanisms such as subsidies to mitigate these potential negative impacts.

Some positive unintended effects of certification are worth briefly mentioning. One example of an unintended positive effect of certification is bringing producers and retailers closer together and shortening the chain of custody. For example, in the case of the Collins Pine Company in California (see Part III, "Geographic Regions" and "Multilateral Agencies and Funders"), the company is selling certified pine shelving directly to The Home

Depot for a higher price than for noncertified timber. Since there are no intermediaries, The Home Depot can sell the shelving to its customers at a lower price.

Other unintended effects have been identified by the Seven Islands Land Company in Maine, which has been certified by Scientific Certification Systems (SCS). These effects include the benefits of independent technical advice: "Therein lies one of the strongest arguments for third-party certification: The landowner can defer questions about the certification to SCS" (McNulty and Cashwell 1995). Other improvements occur in such areas as staff morale: "There has been a significant boost in company morale. Foresters feel rewarded for their efforts. Certification has also been a public relations success. Overall, the value of certification, simply in improvements to this operation, has far outweighed its cost. The well-managed label for use in the marketplace is a bonus" (McNulty and Cashwell 1995).

References

Chris Cox, formerly of Ecological Trading Company, letter to Chris Elliott, August 1993.

da Motta Veiga et al. 1994. "Eco-labeling Schemes in the European Union and Their Impacts on Brazilian Exports." Paper presented at the UNCTAD workshop on Eco-labeling and International Trade, Geneva, June 1994.

Droogsma, W. D., et al. 1991. "Legal Means for Restricting the Impact of Non-Sustainably Produced (Tropical) Timber. Aspects of International and European Law." Center for Environmental Law, University of Amsterdam, Holland. 69 pp.

International Tropical Timber Organization, ITTO. 1994. "Report of the Working Party on Certification of All Timber and Timber Products." ITTO, Yokohama, Japan. 257 pp.

McNulty, J., and J. Cashwell. 1995. "The Land Manager's Perspective on Certification." *Journal of Forestry*, vol. 93, no. 4, 22–25.

Sanford, R. L., et al. 1985. "Amazon Rainforest Fires." *Science* 227: 53–55.

United Nations Conference on Trade and Development, UNCTAD. 1995. "International Cooperation on Eco-labeling and Ecocertification Programs." Report to the ad hoc Working Group on Trade and the Environment, October 1994, UNCTAD, Geneva, Switzerland. 40 pp.

Chapter 12

Certification of Nontimber Forest Products

Virgílio M. Viana, Alan R. Pierce, and Richard Z. Donovan

Historically, the production of nontimber forest products (NTFPs) has been considered marginal to the forest sector. As a result, NTFPs have received very little attention in research and educational programs or in governmental policies. University forestry programs, for example, give very little emphasis to the ecology and management of NTFPs, even in places where they are quite important to the regional economy, as in the Amazon. Likewise, forestry research institutions have traditionally focused almost exclusively on timber products. Similarly, government policies have largely neglected NTFPs, apart from those few that have exceptional market importance.

Over the past decade, the role of NTFPs has begun to be appreciated as a fundamentally important component of sustainable forest management systems. This can be noted from the growing importance of NTFPs in activities of leading international organizations such as the United Nation's Food and Agriculture Organization (FAO), the Center for International Forestry Research (CIFOR), the Centro Agronomico Tropical de Investigacion y Enseñanza (CATIE), and others. Worldwide, in tropical, boreal, and temperate forests, there are thousands of different NTFPs, though the most visible and economically important products include rubber, brazil nuts, medicinal plants, rattan, bamboo, maple syrup, palm heart, berries, and mushrooms.

(See Box 12.1.) In addition to international markets, subsistence uses of NTFPs (such as construction materials for shelter, or food and medicines) are important to many rural communities.

The interest in certification of NTFPs is a newer phenomenon. Although NTFPs have only recently begun to be considered as candidate products for certification, there are many reasons they are important, and there are several management systems and markets that can be regarded as good candidates for certification. This chapter (1) analyzes the importance of NTFPs to sustainable forestry, (2) profiles several promising NTFP initiatives, and (3) discusses some key issues in developing a conceptual framework for the certification of NTFPs.

Importance of NTFPs
to Sustainable Forest Management

Growing interest in NTFPs results from recent research that identifies their importance to (1) forest conservation movements, (2) income generation for rural communities, (3) reduction of adverse environmental impacts to forest ecosystems caused by timber production, and (4) changing the paradigms of conventional forestry.

Forest Conservation

The Brazilian Amazon provides possibly the best known case of a relationship between an NTFP management system and a forest conservation movement. During the 1980s, a precedent-setting alliance was formed between Brazilian rubber tappers, led by the late Chico Mendes, and local and international environmentalists to halt deforestation caused by cattle ranchers

---

Box 12.1

Definition and categories of NTFPs

Nontimber forest products (NTFPs) are "all goods derived from forests of both plant and animal origin other than timber and fuelwood" (FAO 1994:15). NTFPs from plantations and natural forests provide humans with a wide range of useful products, including:

•Nuts, fruits, and greens	•Essential oils	•Spices
•Gums	•Floral trade products	•Medicinal plants
•Resins	•Game	•Latexes
•Dyes	•Fibers	•Insect products

(Allegretti 1990). Cattle ranching resulted in the removal of forests and NTFPs—especially rubber (*Hevea* spp.), Brazil nuts (*Bertholletia excelsa*), and game—upon which rubber tappers depended, making them ready allies with forest conservation NGOs. This alliance resulted in the creation of more than two million hectares (Anderson and Ioris 1992) of extractive reserves in the Amazon. Other similar cases are known in the Philippines, Indonesia, and India, where local communities have depended on natural forests to generate income through local and international markets. In those cases, securing land tenure was an important component of a strategy to promote sustainable management and forest conservation.

In many circumstances, NTFPs have offered alternatives to forest management solely for timber resources. Rattan (*Calamus* spp.) and palm heart (*Euterpe edulis*) are examples of commercially important commodities that complement, to varying degrees, timber extraction. In some cases, the existence of an NTFP will force technicians to make compromises in terms of commercial timber harvesting. In Mexico, rubber tapping customs and the continued, though variable, commercial value of rubber have combined to force very careful consideration of any timber harvesting schemes. NTFPs have thus an important role in maintaining diversity in managed forest ecosystems. (See Box 12.2.)

Income Generation

For many rural communities, NTFPs represent an important source of income. Therefore, any strategy aimed at reconciling the objectives of increas-

---

Box 12.2
Rubber tapping and mahogany in Mexico

In Quintana Roo, Mexico, local *ejido* communities harvest valuable commercial hardwoods and natural rubber from zapote trees (*Manilkara zapota*) on the same land. The customary tradition of rubber tapping has resulted in a constant emphasis on maintaining the zapote tree resource (and the forest), even when international prices for rubber have declined. Maintenance of the zapote resource has also compromised silvicultural techniques for regenerating valuable tropical timber species. For example, mahogany (*Swietenia macrophylla*) requires large openings to stimulate adequate regeneration (Snook 1993; Veríssimo et al. 1995). However, in Mexico, broad-scale adoption of intensive silvicultural methods to regenerate mahogany is not favored by local communities and technicians because of the potential negative impact on the zapote resource.

ing income generation for these communities with forest conservation needs to consider alternatives for increasing productivity and adding value to NTFPs. The socioeconomic importance of NTFPs varies significantly among different regions of the world. In West Kalimantan, for example, NTFPs are very important for traditional forest communities of Dayaks (de Jong 1995). Gunatilake et al. (1993) report that, on average, NTFPs provide greater than 15 percent of the total annual income per household in villages adjacent to the Knuckles National Wilderness Area in Sri Lanka. Reygadas (1991, in Zamora-Martinez and Nieto de Pascual-Pola 1995) has estimated that Mexican mushroom harvesters in the Sierra del Ajusco may earn up to $1,000 per collector during the rainy season. IRRC (1993) estimates that nearly 200,000 forest dwellers earn a living harvesting rattan in Indonesia, and over 500,000 people make their livelihood from rubber extraction in the Amazon. In the Yucatan Peninsula, NTFPs are critical components in Mayan community economies. Similarly, in North America, production of maple syrup from sugar maples (*Acer saccharum*) is an integral economic and social component of rural communities in New England, eastern Canada, and the Lake States region. Nevertheless, the importance of NTFPs to regional or national economies is often underestimated, and there are relatively few statistics on their size and value. (See Table 12.1.)

Environmental Impacts

The production of NTFPs can be expected to produce less severe environmental impacts to forest ecosystems than timber extraction. This is because production of NTFPs often does not involve the harvesting of individual plants or animals, as in the case of Brazil nuts (*Bertholletia excelsa*) and açai fruits (*Euterpe oleracea*) in the Amazon, chicle (*Manilkara sapota*) in Central America, and berry crops (*Rubus* and *Vaccinium* spp.) in North America and Europe. There are, however, other instances where production of NTFPs creates adverse ecological impacts that result in the death of entire plants or individual animals, as the case with palm heart in southeastern Brazil and Costa Rica, ginseng (*Panax quinquefolius*) in North America, and rattan in Asia.

The sustainability of a management system for a NTFP is determined by a number of factors including the population dynamics of a species, the frequency and intensity of its harvest, the characteristics of the forest management system, and the training and education of harvesters. Responses of individual NTFPs to harvesting pressures are variable. In eastern North America, over-harvesting of edible fiddlehead ferns (*Matteuccia struthiopteris*) has precipitated a decline in the number, size, and flavor of ferns taken

Table 12.1
Market value for some nontimber forest products

Country	Product(s)	Value(U.S.$)	Year	Source
Brazil	Brazil nut exports	50 million/yr.	1990	Viana et al. 1994
China	All NTFP exports	340 million/yr.	1990	FAO 1994
Ghana	Bushmeat	202,000/ Single market day in Accra	1985	FAO 1994
India	Sandalwood oil exports	5.4 million/yr.	1991–92	FAO 1994
	Tendu leaves	200 million/yr.	1991	FAO 1994
Indonesia	Rattan furniture exports	247 million/yr.	1992	FAO 1995
Japan	All NTFPs	3.6 billion/yr.	1991	FAO 1994
Sudan	Gum Arabic exports	47 million/yr.	1988–89	FAO 1994
U.S.A.	Maple syrup	32.2 million/yr.	1994	USDA 1995
U.S.A.	Wild mushrooms (Idaho, Oregon, Washington only)	20.2 million/yr.	1992	Schlosser and Blatner 1995

from frequently collected sites (von Aderkas 1984). Nault and Gagnon (1993) found that wild leeks (*Allium tricoccum*) in Canada showed symptoms of population decline at simulated harvest rates of only 5 to 15 percent, thus explaining reported local extinctions of the plant. Conversely, Kardell (1980) concluded that annual harvests of forest berries and edible wild mushrooms in Sweden was probably 10 times lower than the potential sustainable harvest level.

There are, however, few quantitative analyses on the sustainability of NTFPs. A noteworthy exception is the work of Peters (1990), who used matrix models to analyze the sustainability of *Grias peruviana* fruit harvests in Western Amazonia. The maximum intensity of sustainable harvest for this species was determined to be about 80 percent of the fruit crop.

Paradigm Changes

Forest management for NTFPs has an important role in changing the paradigms of forestry and conservation. Historically, the focus of university-trained forest managers was almost exclusively directed to timber products.

As the concept of ecosystem management has become widely discussed in the academic and professional world (SAF 1993), nontimber forest products are receiving increasing attention. Management for multiple products (timber and nontimber) presents opportunities to "maintain the complex processes, pathways, and interdependencies of forest ecosystems and keep them functioning well over long periods of time" (SAF 1993: xvi).

Maintenance of forest ecosystem integrity increases the probabilities for immediate and future ecological and economic benefits (SAF 1993). Managing for NTFPs will require changing the content of forestry curricula and research to include information on the population biology and management of forest herbs, shrubs, palms, fungi, and vines.

In the process of expanding the breadth of mainstream forestry, special attention should be given to management systems developed by traditional forest communities (Padoch and Peters 1993). In regions like the Amazon or Kalimantan, where traditional populations have developed appropriate technologies for millennia, rethinking the paradigms of forestry requires necessarily an understanding of those management systems by formal forestry science.

Conservation strategies have often neglected the role of traditional populations in the management of NTFPs (Diegues 1994). The understanding and recognition of the value of traditional forest management systems by mainstream forestry will, in turn, require a new ethos as it implies respect for societies and cultures that have long been considered "primitive" and "backward" (Posey 1992). This new ethos is an important component of a broad strategy for the conservation of cultural diversity, an important—and often forgotten—component of sustainable development principles. Sustainable management of NTFPs also provides an opportunity to address critical challenges posed by the "marriage" between conservation and sustainability (Redford and Sanderson 1992). If forest products are being sustainably harvested and forests are being protected against severe human disturbance and conversion to alternate uses, forests managed for NTFPs may provide an important complement to nature preserves.

Initiatives Toward Sustainable Forest Management

Initiatives toward sustainable management of NTFPs are taking place in many forested regions of the world. Some of these initiatives represent new development projects supported by NGOs, development agencies, and foundations, while others represent initiatives by commercial interests, such as pharmaceutical companies. In addition, and most important, there are those management systems that simply represent old practices of natural forest management that are now being "discovered" by Western science.

A number of cases of NTFP management initiatives by traditional communities have only recently been documented (Clay 1988). Padoch and Peters (1993) and de Jong (1995), for example, studied the structure and management of *tembawangs* (fruit forest gardens) in Southeast Asia and concluded that they had a very high species diversity and other ecosystem characteristics that make them quite similar to old-growth forest patches. Some *tembawangs* may have more than 90 percent of the number of species and 80 percent of the basal area of old-growth forest patches (de Jong 1995). These management systems have often been neglected by land-use planners and research programs, despite their importance for both income generation and forest conservation.

Efforts to develop sustainable production of NTFPs in both natural forest and plantation management systems are supported by a number of industries, NGOs, and governments. Examples include rattan plantations by Innoprise Sbn. in Sabah, Malaysia; agroforestry systems for black pepper (*Piper nigrum*) production in combination with *Gliricidia sepium*, and vanilla (*Vanilla planifolia*) production in combination with *Terminalia ivorensis*, both in Costa Rica; and palm heart plantations in the Brazilian Atlantic rainforest. Some agroforestry schemes are stand-alone commercial projects, having little or nothing to do with forest management. In other cases, such as the KORUP project in Cameroon and the ANAI and BOSCOSA projects in Costa Rica, agroforestry schemes are implemented with the specific aim of complementing or reinforcing the development of a "forestry culture," or *cultura forestal* (Donovan 1994). Agroforestry activities are specifically incorporated in projects that also include natural forest management and commercial timber extraction.

Certification Initiatives

Certification initiatives for NTFPs developed later than those for timber certification programs. Part of the reason is that there was greater concern from the public about the unsustainability of wood products than about NTFPs. However, because of their relative lower environmental impact and higher social benefits, certification initiatives of NTFPs are developing quite rapidly. Economic motivations for the development of certification initiatives for NTFPs include (1) higher market share, especially in industrialized urban centers; (2) higher premium prices, especially for "gourmet" products; and (3) higher per unit value of NTFPs compared to timber.

Environmental motivations include (1) unsustainable harvests of many NTFPs, including endangered species; (2) environmental impacts (e.g., wildlife impoverishment, soil erosion, pesticide contamination, and so on) of intensive production systems replacing wild population-based management systems; and (3) environment benefits of promotion of diversified forest management (e.g., biodiversity conservation). Socially driven motivations include (1) benefits of increased revenues to producers and (2) reduced economic risk resulting from diversified production systems of NTFPs.

Certifiers of NTFPs need to consider that production systems may range from simple gathering of products in natural forests (such as Brazil nuts in Amazonia) to intensively managed plantations (such as palm hearts in Central America). There are also intermediate cases with intensive management of natural forests that include planting and thinning practices (for example, *tembawang* fruit gardens in Kalimantan).

A group of Brazilian NGOs, research and government institutions, and the executive secretariat of IMAFLORA (Instituto de Manejo e Certificacao Florestal e Agricola), with support from Cultural Survival Enterprises, the Agroextrativist Cooperative of Xapuri (CAEX), and Rainforest Alliance, are developing NTFP certification standards. Initial emphasis has been focused on Brazil nuts and natural rubber production in Brazil and Bolivia. These standards will be used as a basis for other certification initiatives in Mexico, Guatemala, and Costa Rica, where they may be expanded to cover other products such as palm heart, chicle, and pine resin. (See Box 12.3.)

In the Western United States, a number of regional NGOs and forest products marketing specialists are evaluating the certification potential of various NTFPs including mushrooms, evergreen huckleberry (*Vaccinium ovatum*), salal (*Gaultheria shallon*), Christmas boughs (various species of conifers), yerba santa (*Eriodictyon califoricum*), sword fern (*Polystichum munitum*), beargrass (*Xerophyllum tenax*), and Oregon grape (*Mahonia aquifolium*). One organization in particular, the Rogue Institute for Ecology and Economy (RIEE), is focusing its research efforts on the education and training of NTFP harvesters (Everson, personal communication 1995). Once the social assessment of harvesters is completed, RIEE plans to conduct ecological research on individual species, hoping to establish comprehensive NTFP harvest standards.

In the Northeastern United States, labeling of maple syrup made from sugar maple has been regulated by the State of Vermont's Department of Agriculture for 45 years (State of Vermont 1949). This certification system provides an interesting example of a state government labeling program that

Box 12.3

Pine resin tapping in Michoacan, Mexico

Resin tapping is a traditional production activity in many rural communities in Mexico. The Nuevo San Juan cooperative in Michoacan state manages roughly 20,000 hectares of forest, with a commercial production of roughly 100,000 cubic meters of wood and 1000 tons of resin (Saucedo, personal communication 1995). Three native pine species are tapped, including *Pinus michoacana*, *Pinus pseudostrobus*, and *Pinus montezumi*, and the resin is sold nationally and internationally as a base for manufacturing commercial paints.

Tree regeneration is mostly natural, though some 2 million seedlings are planted annually (Saucedo 1995) in areas where natural regeneration is sparse or otherwise problematical. Forestry activities are managed by a cooperative utilizing a technical team of four foresters, four forest technicians, and three *"practicos"* (a person who specializes in forest management but has received no formal academic training). Resin tapping is a centuries-old activity that has contributed a crucial source of revenue in both modern (since 1970) and traditional times. According to interviews, resin tapping provides a relatively low but consistent level of income and has served as an incentive to keep trees on the land over the years. Community members (*"comuneros"*) get roughly 2.5 kilos per *"cara"* (the cut face on a tree) of resin, which currently sells for roughly 225 pesos per ton (U.S.$44)(Saucedo 1995). Resin cooperatives in Nuevo San Juan currently produce about 600 to 1000 tons per year, but believe they have the potential to produce up to 2000 tons per year (Saucedo 1995).

Comuneros are now experimenting with *"resinacion intensiva,"* a technique that involves cutting up to eight small caras per tree; in the past, only one to two *caras* were made on each tree. *Comuneros* perform the *resinacion intensiva* only when they are about to harvest a tree, thus having minimal impact on regeneration or the quality of the trees that will continue to grow in the forest. Some good-quality wood is lost due to both types of resin tapping because the activity makes the bottom 1.5 to 2 meters of the tree unmerchantable. The resin tapping cooperatives are willing to forgo this loss of commercial timber because of the overall long-term benefits that resin tapping and tree conservation provide to the *comuneros*.

has (1) ensured quality control of a forest product and (2) resulted in a willingness on the part of consumers to pay premium prices for labeled products. Although this certification does not objectively evaluate the environmental quality of sugar maple stands, the "green" aspect of the product is used in its marketing, based largely on the "green" image of the State of Vermont. (See Box 12.4.)

Box 12.4

Maple syrup production in Vermont, USA

U.S. maple syrup production in 1995 was 4.1 million liters, estimated to be worth over $25 million (USDA 1995). The State of Vermont is the country's leading producer of maple syrup, supplying approximately one-third of total national production. There are nearly 2,500 maple producers in Vermont. Large-scale, intensive sugaring operations can gross as much as $2,500 per hectare (Sendak, personal communication 1994).

The making of maple syrup is a centuries-old art originally taught to European settlers by Native Americans (Nearing and Nearing 1970). Maple stands ("sugarbushes") are created by thinning undesirable species and lowering basal area to create wide-crowned maples, which typically produce more sap than narrow-crowned trees. Maple sap is harvested in late winter and early spring when temperatures cause sap flow to be most abundant. Tap holes are drilled into the maple trees and sap is collected in buckets or by a network of plastic tubing. Excess water is evaporated from the sap to make the final product.

Maple syrup is Vermont's flagship specialty product, and the product is inextricably bound to the image of the state. Maple syrup has been graded in Vermont by the state's Department of Agriculture since the late 1920s on a voluntary basis, and since 1950 on a compulsory basis. The state's sugar makers have strongly supported the grading and labeling system since its conception (State of Vermont 1952). Maple syrup is graded by color, flavor, and density.

The State of Vermont issues "Vermont Seal of Quality" labels to agricultural products (maple is considered an agricultural product) and "Vermont Makes It Special" labels to nonagricultural products. To qualify for such labels, producers must meet sanitary requirements, allow spot inspections of their products and facilities, and use 100 percent Vermont products or add significant value to imported non-Vermont products.

Vermont represents a case of state-level governmental "eco-labeling" of products based on product standards and public image, not on direct assessment of ecological standards. A marketing report commissioned for the state discovered that consumers expect to pay more for Vermont products because they are perceived to be superior/gourmet goods (Gourley and Myers 1986). Producers and the state feel that the state-issued labels promote local jobs and increase market share sales by capitalizing on the subjective image consumers have of Vermont as a "green," rural, scenic state that produces high-quality products. About 150 large-scale producers of maple products use the "Vermont Seal of Quality" and "Vermont Makes It Special" labels when exporting their products out of the state or out of the country (Marckres, personal communication 1995). The higher prices paid for Vermont goods bearing labels of origin or in the case of maple syrup, labels of quality assurance, demonstrate a case where governmental labeling of products is trusted and economically attractive.

Vermont has been able to parlay its green image (both real and perceived) into profits. Consumers of pure Vermont maple syrup likely believe their purchase supports a "green" product as well as a "green" state. After all, maple tapping does not kill maple trees, keeps land covered in forest, provides beautiful scenery in the fall when the maple leaves turn color, offers carbon sink storage, and supports rural economies. Certifiers of maple products may choose to highlight the potentially adverse aspects of sugarbush management (soil compaction, disease and insect susceptibility, localized loss of biodiversity, impact on wildlife habitat, nutrient depletion). However, it is likely that consumers will continue to believe that maple syrup is a pure, sustainable forest product unless these potentially negative effects of sugaring are publicized.

Key Issues in Certification of NTFPs

Certification of NTFPs faces many challenges. Here we discuss some key issues that need to be addressed in the development of certification programs for NTFPs.

Intensity of Management. The intensification of management systems of NTFPs to increase competitiveness with alternative land uses can result in conflicts between economic returns and environmental quality. Increasing the density of desirable species often results in lower biodiversity. In the case of forests managed by Dayak communities in Kalimantan, for example, intensively managed *tembawangs* have resulted in loss of more than 50 percent of tree species (de Jong 1995). The same is true for maple sugarbushes in North America, where intensive management practices reduce tree species diversity and may result in conflicts with other forest uses and values, such as wildlife habitat protection (Beattie et al. 1993). A question that needs to be addressed from a scientific and political perspective is the extent to which reductions of forest species diversity are compatible with the principles of sustainable forestry. Is there a minimum species diversity to be maintained in a managed forest? Or, alternatively, should species diversity be maintained at the landscape level, giving more flexibility to natural forest managers to intensify production while not favoring conversion to plantations?

Natural Forest NTFP Production Systems Versus Plantation Systems. Several authors have questioned the productive potential of NTFPs in natural forests based on claims that domestication would eventually drive them out of competition (Homma 1992). While this may be true for some products such as rubber and cacao (*Theobroma cacao*), it does not seem appropriate to make a generalization for all NTFPs. There are many examples of products that are more competitive in wild or semi-wild conditions than in domesticated plantations. An example is ginseng (*Panax* spp.)production, which is much more valued from natural stands than from plantations (Box 12.5). Other examples include some ornamental plants that are difficult to cultivate, vines, and so on.

The agricultural paradigm of moving production systems of NTFPs from natural stands to plantations needs to be analyzed critically. There are many instances where traditional peoples developed forest management techniques that resulted in semi-wild populations of trees and animals with higher productivity, broader environmental benefits, and greater appropriateness to low capital availability, such as the Dayaks in Indonesia. (See earlier sections of this chapter.) The old agricultural paradigm of promoting domestication of NTFPs needs to be reconsidered in light of recent studies in the fields of resource economics, ecological economics, and ecological anthropology, which reveal that traditional agroforestry systems provide broader socioeconomic and environmental advantages than plantations.

Box 12.5

Ginseng production in the United States of America

The United States is the world's third largest producer of ginseng, behind South Korea and China respectively. U.S. exports of American ginseng (*Panax quinquefolius*), an herb used in traditional Chinese medicine, totaled nearly 1.1 million kilograms in 1994, and was valued at over $75 million (U.S. Dept. of Comm. 1995). Nearly 90 percent of the entire U.S. production of ginseng is exported (Mater Engineering 1993). In 1994 more than 80 percent of the U.S. ginseng exports were shipped to Hong Kong. North American ginseng is currently listed under Appendix II of the Convention for International Trade in Endangered Species of Wild Fauna and Flora (CITES), and all cultivated and wild ginseng must be certified as to origin before entering the international marketplace.

There are four distinct classes of North American ginseng roots: wild, semi-wild, woods-cultivated, and artificial-shade-cultivated. Wild ginseng grows in scattered populations throughout the eastern hardwood forests of North America. Wild ginseng is collected by rural inhabitants who receive as much as $320 per pound for premium roots (Mater Engineering 1993). Semi-wild ginseng is cultivated beneath natural forest canopy in low-density "enrichment planting" plots and requires limited tending. Prices paid for semi-wild ginseng range from $60 to $300 per pound (Persons 1994). Woods-grown ginseng is more densely planted than semi-wild ginseng and often entails heavy modification of the forest understory using tractors or rototillers along with herbicide, fungicide, and fertilizer treatments. Prices paid for woods-grown ginseng range from $60 to $135 per pound (Persons 1994). Artificial-shade-grown ginseng is cultivated in labor-and-capital-intensive plantations in open fields beneath wooden or plastic shading material. Average prices paid for artificial-shade-grown ginseng fell to just under $24 per pound in 1994 (U.S. Dept. of Comm. 1995).

Prices paid for ginseng are determined by a root's age, size, shape, color, and texture. Cultivated roots are easily differentiated from semi-wild and wild roots by appearance alone. The significant difference in prices paid for artificial-shade-grown ginseng, semi-wild ginseng, and wild ginseng represents a clear example of a NTFP that is more highly valued when produced in wild or semi-wild conditions than in a plantation. Asian ginseng (*Panax ginseng*), for example, is extremely rare in the wild, and single wild roots of the plant have sold for as much as $20,000 (Mater Engineering 1993).

Certification has the potential to provide greater economic incentives for ginseng production in wild and semi-wild conditions and could result in the internalization of environmental values inherent in these production systems (conservation of biodiversity and soils, carbon storage, no use of pesticides). However, the willingness of ginseng consumers to pay premium prices for such environmental values remains unknown. Control of illegal ginseng harvesting will prove to be a challenge for certifiers. Yet, if properly implemented, certification could conceivably complement the aims and objectives of CITES.

Poor Scientific Basis for NTFP Management. Compared to timber products, NTFPs have been relatively underresearched. For many NTFP species, knowledge of the basic aspects of their population biology remains unknown or incomplete. This lack of basic ecological information holds true even for widely marketed NTFPs, and thus presents an additional challenge to NTFP certification. For example, until very recently there was very little quantitative information on the demography of Brazil nut populations. Contrary to previous understanding, regeneration of Brazil nut has been found to be abundant (Viana et al. 1994). This type of information is critical not only for certification but also for developing alternatives to increase the productivity of NTFPs. Low productivity of NTFPs can be related, to a significant extent, to the scanty research on the biology and management of those species.

Sustainability Analysis of NTFPs. Certifying NTFP management systems requires different approaches to the sustainability analyses used for timber production systems. Population dynamics models can be quite precise in determining sustainable harvest rates (Figure 12.1). Peters (1990), for example, used matrix models to analyze the sustainability of *Grias peruviana* fruit harvests in Western Amazon. However, quantitative analyses predicting sustain-

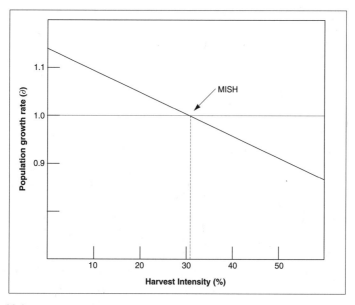

Figure 12.1
Population growth rate against harvest intensity, indicating the theoretical maximum level of sustainable harvest of NTFPs. MISH is the maximum intensive sustainable harvest.

able production levels of NTFPs are scarce. Simpler analyses of the population structure of plants can be quite informative for rapid assessments (Hall and Bawa 1993; Peters 1994). In conducting such surveys, it is vital to devise appropriate sampling strategies, since regeneration of many NTFPs plant species is often patchy, and inappropriate sampling may give misleading results.

In addition to management-oriented research, there is also a paucity of information on processing and marketing of most NTFPs. It is no surprise that waste levels are often high and marketing of NTFPs is poor. Similarly, there is relatively little information on socioeconomic aspects of NTFPs production. There is a pressing need for more research on the biology, management, processing, and marketing of NTFPs.

Other NTFP Labeling Initiatives. Certification of NTFPs sometimes overlaps with certification of organic products. IFOAM (International Federation of Organic Agriculture Movements)-accredited certifiers are implementing standards for maple syrup, commercially cultivated mushrooms, and wild-harvested edible and medicinal plants, some of which are forest-dwelling species. The Organic Crop Improvement Association (OCIA), the world's largest international certifying body of organic products, requires that member chapters which produce and sell NTFPs submit themselves to regular inspections and audits. OCIA-member NTFP producers must prove that all wild-harvested plants come from pesticide-free forests under active management plans. OCIA criterion 4.6.2 for wild plants also states that: "harvest (of wild plants) must de done in a way that maintains the natural balance of the ecosystem" (OCIA 1995: 19). While specific standards for maple syrup and commercial mushroom production exist, no harvest standards for individual species of wild-collected plants are in place (although local chapters are free to draft specific NTFP standards as long as they meet with OCIA approval). There is, however, a need to differentiate the certification of products that are purely organic from those that are also environmentally and socially sound. Consumers need to be able to make this distinction.

Public Education and Marketing of NTFPs. The marketing of certified NTFPs presents different challenges than that of timber products. The marketing of NTFPs seems to meet with fairly positive public reaction due to the public perception that one can harvest NTFPs and still "keep the forest." This is both philosophically and aesthetically pleasing, though not necessarily true. This general perception should facilitate the development of certified NTFP markets. However, public reaction to the need for certification of NTFPs is mixed. For example, in 1994, Rainforest Alliance conducted interviews with major producers, importers, and retailers of rattan products for the U.S. market. These interviews suggested that the market for a certified rattan product would be minimal unless a major public education effort took place to inform consumers of the negative impacts of many nonsustainable rattan sources. The same scenario is likely to be true for other NTFPs, although current quantitative information does not allow a definitive analysis of this issue.

Timber certification has been able to capitalize on massive public education efforts that have dramatized the plight of the rainforest. Without similar education efforts and without the involvement of key commercial interests, certification of NTFPs may be perceived as a marginal issue. Thus, public education is a key challenge for the commercial success of certification of NTFPs.

Equity Issues. As commercialization of NTFPs has increased, it is not always clear if the economic benefits have been equitably shared with harvesters, or that adequate resources have been invested in improving the ecological aspects of NTFP management. Some companies, like Cultural Survival Enterprises, the Body Shop, and Ben & Jerry's, have made commitments to equitable and ecologically sound commercial NTFP development. Yet even for such well-meaning companies, mechanisms are not always in place for ensuring that local people are adequately compensated, or that environmental controls on production techniques are in place.

Subsistence Communities and Cultural Integrity. Forest-dependent subsistence economies often do not necessarily value NTFPs in monetary terms. Researchers have found that NTFPs are reciprocally exchanged in subsistence communities from Alaska (Glass et al. 1989) to Amazonia (Smith 1995) as gifts. Profit motivation is often foreign to subsistence communities (Smith 1995), and the introduction of such a concept via NTFP marketing may prove perilous to traditional social structures. Conflict often erupts when highly valued NTFPs lure profit seekers into forests traditionally used by subsistence cultures. Protection of the traditional social structure of subsistence communities, and protection of the products relied upon by these cultures from exploitation by outside forces, pose major challenges for certification.

Chain of Custody for NTFPs. NTFPs create special problems for certification from a chain-of-custody perspective. NTFPs are often collected from a widely dispersed resource base; therefore, many opportunities exist, in theory, to mix nuts or rubber from a certified production area with product from noncertified operations. There are typically many steps in the processing of NTFPs that provide opportunities for unscrupulous "contamination" of cer-

tified products. Certifiers of NTFPs will need to give increasing attention to
the chain-of-custody issue.

Conclusions

Although certification of NTFPs is still in an embryonic state, there seem to
be many cases where it is likely to be successfully developed and implemented. This is particularly true for those products aimed at industrialized
urban centers, especially for "gourmet" products. There are also good
prospects for compatibility between certification of NTFPs and conservation
efforts, particularly in forested areas that are inhabited by traditional populations.

Certification of NTFPs has many challenges ahead, including the incorporation of concepts of landscape ecology, particularly in intensively cultivated systems; development of simpler tools for sustainability analysis as well
as tackling other research gaps; differentiation from organic certification;
more equitable distribution of certification revenues; and marketing.

Finally, the importance of certification of NTFPs comes from its relevance
to forest conservation movements and its potential to generate income for
rural communities. Due to their potential positive environmental and socioeconomic benefits, production of certified NTFPs may play an important
role in changing the paradigms of mainstream forestry.

References

Allegretti, M. 1990. Extractive reserves: an alternative for reconciling development
    and environmental conservation in Amazonia. Pages 252–264 in A. Anderson,
    ed. *Alternatives to deforestation: steps toward sustainable use of the Amazon rain for-
    est*. New York: Columbia University Press.

Anderson, A., and E. Ioris. 1992. "Valuing the rain forest: economic strategies by
    small-scale forest extractivists in the Amazon estuary." *Human Ecology* 20 (3),
    337–369.

Beattie, M., C. Thompson, and L. Levine. 1993. *Working with your woodland: a
    landowner's guide*, revised edition. University Press of New England, Hanover,
    N.H. 279 pp.

Clay, J. 1988. "Indigenous Peoples and Tropical Forests." Cultural Survival Report no.
    27. Cultural Survival, Cambridge, MA. 116 pp.

de Jong, W. 1995. Recreating the forest: successful examples of ethnoconservation
    among Dayak groups in central West Kalimantan. Pages 295–304 in Ø. Sand-

bukt, ed. *Management of tropical forests: toward an integrated perspective.* Center for Development and the Environment. Oslo, Norway. 381 pp.

Diegues, A.C. 1994. O Mito da Natureza Intocada. NUPAUB/Universidade de São Paulo. São Paulo, Brazil.

Donovan, R. 1994. BOSCOSA: Forest conservation and management through local institutions (Costa Rica). Pages 215–233 in D. Western, R. Wright, and S. Strum, eds. *Natural connections: perspectives in community based conservation.* Island Press, Washington, D.C. 581 pp.

Everson, B. 1995. Coordinator for special forest products project, Rogue Institute of Ecology and Economy. Personal communication.

FAO. 1994. *Non-Wood News* vol. 1, March 1994. Food and Agriculture Organization of the United Nations, Non-Wood Products and Energy Branch, Forest Products Division, Rome, Italy. 47 pp.

FAO. 1995. *Non-Wood News* vol. 2, March 1995. Food and Agriculture Organization of the United Nations, Non-Wood Products and Energy Branch, Forest Products Division, Rome, Italy. 79 pp.

Glass, R., R. Muth, and R. Flewelling. 1989. "Subsistence as a component of the mixed economic base in a modernizing community." U.S. Department of Agriculture, research paper NE-638, Northeast Forest Experiment Station, Radnor, PA. 9 pp.

Gourley, L., and N. Myers. 1986. "The qualitative investigation of the image of the State of Vermont." Research paper prepared for the State of Vermont Agency of Development and Community Affairs. Marketing Dynamics Group, Inc. Burlington, VT.

Gunatilake, H., D. Senaratine, and P. Abeygunawardena. 1993. Role of nontimber forest products in the economy of peripheral communities of Knuckles National Wilderness Area of Sri Lanka: a farming systems approach. *Economic Botany* 47 (3), 275–281.

Hall, P., and K. Bawa. 1993. Methods to assess the impact of extraction of nontimber tropical forest products on plant populations. *Economic Botany* 47 (3), 234–247.

Homma. 1992. The dynamics of extraction in Amazonia: a historical perspective. Pages 23–31 in D. Nepstad and S. Schwartzman, eds. Nontimber products from tropical forests: evaluation of a conservation and development strategy. *Advances in Economic Botany* vol. 9. New York Botanical Garden, N.Y. 164 pp.

IRRC. 1993. Nontimber forest products and ecotourism. Pages 165–192 in C. MacKerron and D. Cogan, eds. *Business in the rain forests: corporations, deforestation and sustainability.* Investor Responsibility Research Center, Washington, D.C.

Kardell, L. 1980. "Forest berries and mushrooms—an endangered resource?" *Ambio* 9 (5), 241–247.

Laird, S. 1993. "The natural management of tropical forests for timber and nontimber products." Master's dissertation, Oxford Forestry Institute, Oxford University.

Marckres, H. 1995. Agricultural products specialist, consumer assurance section, Vermont Department of Agriculture. Personal communication.

Mater Engineering, Ltd. 1993. "Minnesota special forest products project: final report." Mater Engineering, Ltd., Corvallis, OR. 100 pp.

Nault, A., and D. Gagnon. 1993. "Ramet demography of *Allium tricoccum*, a spring ephemeral, perennial forest herb." *Journal of Ecology* 81 (1), 101–119.

Nearing, H., and S. Nearing. 1970. *The maple sugar book*. Schocken Books, New York. 273 pp.

OCIA. 1995. "1995 International bylaws," as revised February 1995. OCIA, Bellefontaine, OH.

Padoch, C., and C. Peters. 1993. "Managed forest gardens in West Kalimantan, Indonesia." In C.S. Potter et al., eds. *Perspectives on biodiversity: case studies of genetic resource conservation and development*. Washington, D.C.: AAAS.

Persons, W.S. 1994. *American ginseng: green gold*. Bright Mountain Books, Inc., Asheville, N.C. 203 pp.

Peters, C. 1990. "Population ecology and management of forest fruit trees in Peruvian Amazonia." Pages 86–98 in A. Anderson, ed. *Alternatives to Deforestation: Steps Toward Sustainable Use of the Amazon Rain Forest*. Columbia University Press, New York.

Peters, C. 1994. "Sustainable harvest of nontimber plant resources in tropical moist forest: an ecological primer." Biodiversity Support Program/WWF. Washington, D.C.

Posey, D. 1992. Interpreting and applying the "reality" of indigenous concepts: what is necessary to learn from the natives. Pages 21–34 in K. Redford and C. Padoch, eds. *Conservation of neotropical forests*. Columbia University Press, New York.

Redford, K. and S. Sanderson. 1992. The brief, barren marriage of biodiversity and sustainability? *Bulletin of the Ecological Society of America* 73(1): 36–39.

SAF. 1993. Task force report on sustaining long-term forest health and productivity. Society of American Foresters, Bethesda, MD. 83 pp.

Saucedo Soto, A. 1995. Director Tecnico Forestal, Nuevo San Juan Cooperative. Personal communication.

Schlosser, W., and K. Blatner. 1995. The wild edible mushroom industry of Washington, Oregon, and Idaho: a 1992 survey. *Journal of Forestry* 93 (3), 31–36.

Sendak, P. 1994. USDA Forest Service, Northeastern Forest Experiment Station, Durham, N.H., personal communication, September 1994.

Smith, R. 1995. The gift that wounds: charity, the gift economy, and social solidarity in indigenous Amazonia. Conference paper presented at: Forest ecosystems in the Americas: community management and sustainability, February 3-4, 1995, University of Wisconsin, Madison, WI. 33 pp.

Snook, L. 1993. Stand dynamics of mahogany (*Swietenia macrophylla* King) and associated species after fire and hurricane in the tropical forests of the Yucatan Peninsula, Mexico. Ph.D. dissertation, School of Forestry and Environmental Studies, Yale University.

State of Vermont. 1949. An act to regulate the sale of maple syrup in the state. Acts and resolves passed by the general assembly of the State of Vermont at the 40th biennial session. Montpelier, VT.

State of Vermont. 1952. Report of the Vermont maple sugar maker's association, Inc. Pages 78–82 in Agriculture of Vermont: 26th biennial report of the commissioner of Agriculture of the State of Vermont, 1951-52. Montpelier, VT.

U.S. Department of Agriculture. 1995. Maple Syrup. New England Agricultural Statistics Service, June 14, 1995. Concord, N.H. 4 pp.

U.S. Department of Commerce, Bureau of the Census. 1995. Merchandise trade—imports and exports by commodity: ginseng roots, cultivated and ginseng roots, wild (CD-ROM). Available: national trade data bank—the export connection (R) program: U.S. trade information, 1995.

Veríssimo, A., P. Barreto, R. Tariffa, and C. Uhl. 1995. Extraction of a high-value natural resource in Amazonia: the case of mahogany. *Forest Ecology and Management*. In press.

Viana, V., R. Mello, L. Morais, and N. Mendes. 1994. Ecology and management of Brazil nut populations in extractive reserves in Xapuri, Acre. University of São Paulo, Brazil.

von Aderkas, P. 1984. Economic history of ostrich fern, *Matteuccia struthiopteris*, the edible fiddlehead. *Economic Botany* 38 (1), 14–23.

Zamora-Martinez, M., and C. Nieto de Pascual-Pola. 1995. Natural production of wild edible mushrooms in the southwestern rural territory of Mexico City, Mexico. *Forest Ecology and Management* 72, 13–20.

Chapter 13

Research Needs and Information Gaps

Francis E. Putz

Forest product certification provides a mechanism for rewarding forest managers who treat their resource base in ecologically, silviculturally, socially, and economically sound manners. To promote conservation through wise resource stewardship, foresters must continue to refine definitions of "sound" forestry and further develop methods for evaluating forest management practices (Boyle and Sayer 1995). In this chapter I suggest some of the types of applied research needed as a basis for these evaluation methods as well as some of the more fundamental research needs of environmentally and socially concerned forest managers. I am a tropical forest ecologist/silviculturalist and naturally focus this chapter on biophysical issues related to certification in low-latitude forests. I also attempt to identify some research needs in forests outside of the tropics and endeavor to point out at least a few examples of research on social and economic issues needed in support of certification. Putz and Viana (1996) discussed some other research needs as well as the importance of identifying meaningful and easily assessed indicators of good forestry. Plantation management is treated in more detail elsewhere (Evans 1984; Brown and Nambiar 1995).

Given the tremendous variety of forests in the world and the inadequacies of our understanding of their silviculture, ecology, and roles in human society, it should be no surprise that forest product certification is and will continue to be an "adaptive" process. As we learn more about the species for which forests are managed and the nontarget species and ecological processes that are influenced by management activities, the definition of sound management and associated certification standards will need to evolve. Even in parts of the world with long histories of forestry research, such as in Scandinavia and portions of North America, what was long accepted as good forestry has changed dramatically during the last few years. Where forests are orders of magnitude more species-rich and substantially less well known, as in many parts of the tropics, dramatic changes in accepted forest management practices are also likely. The objective of this chapter is to outline some of the types of research that will help forest managers and forest certifying agents identify changes in forest management practices necessary to more rapidly approach the elusive goal of sustainability.

Research priorities suggested in this chapter are divided somewhat arbitrarily into those related to (1) timber harvesting, (2) effects of silvicultural treatments on commercial species, (3) silvicultural effects on other (noncommercial) species and ecosystem processes, and (4) social issues related to forest product certification. In each section I attempt to address the issues at scales including individual species populations, stands (patches of vegetation that are homogeneous enough to respond uniformly to silvicultural treatments), entire forests, and landscapes.

Research on Timber Harvesting

Properly planned and conscientiously implemented timber harvesting operations can function as very effective silvicultural treatments. Unfortunately, environmentally destructive logging practices are the rule rather than the exception in many forests, especially in the tropics. Dramatic evidence of the potential to reduce damage to soil and advanced regeneration of commercially important species in tropical forests by planning skid trails and roads, pre-felling vine cutting, directional felling, and right-sizing and careful utilization of yarding equipment is accumulating from a developing network of "reduced-impact logging" studies in the humid tropics (Veríssimo et al. 1992; Cedergren et al. 1994; Pinard et al. 1995). While general environmental guidelines for logging, applicable worldwide, are provided by the *FAO Model Code of Forest Harvesting Practices* (Dykstra and Heinrich 1995), much research remains to be done. Reducing residual stand and site damage due to logging should remain a priority (Froehlich and McNabb 1984), but additional research is also needed on methods for restoring the productivity of

compacted or otherwise degraded soils and recovery of areas with logging-in-
duced weed infestations.

One of the challenges of trying to identify research priorities related to
timber harvesting is the tremendous range of technological approaches to log
yarding. Where trees are felled with axes, converted on the spot to lumber to
in pit saws, and carried out of the forest on peoples' backs, research on the er-
gonomics of trump lines and on ax handle flexibility might be appropriate.
Where skyline yarding systems could be used, in contrast, engineering studies on carriage design and intermediate spar selection might be more appropriate. Over the entire range of logging methods, however, economic analyses are needed that include the full range of costs and benefits to assure that
certification guidelines are neither unduly onerous nor overly lenient. For example, in situ conversion of logs into lumber using chainsaws simultaneously
represents a solution to the problem of yarding large logs, an approach to
added value processing available to relatively poor people, a substantial loss of
timber due to wide saw kerfs, and a major challenge for anyone trying to
monitor logging activities. Researchers need to inform policy makers and certifiers about different types of "walk about" sawmills, including their relative
social benefits, environmental costs, and silvicultural challenges.

Major environmental problems associated with logging often result from
the use of inappropriate yarding equipment. In particular, due to their ready
availability and ability to skid large logs up steep slopes on wet ground,
crawler tractors (bulldozers) designed to move earth and build roads are often
also used to yard logs. The damage wrought by bulldozers can be substantially reduced, however, by limiting their use to preplanned skid trails, restricting blade use, and maximizing winching distances (Pinard and Putz
1996). Further reductions in bulldozer damage are definitely possible but will
require research conducted in selectively logged forests under wet tropical
conditions. For example, studies are needed on methods to increase the mechanical efficiency of winching (e.g., with integrated arches) and skidding
(e.g., with skidding pans), as well as on the costs and benefits of wider bulldozer tracks and smaller blades (Jonsson and Lindgren 1990). The effects of
variable tire inflation systems for log-hauling vehicles (Bradley 1993) on sed-
iment loads in logged catchments, road maintenance, and hauling costs
should also be investigated in tropical forests.

Where forests are selectively logged, damage to some potential crop trees
is inevitable. The amount of damage that is deemed acceptable will vary with
forest type and silvicultural approach. Note that there is no damage to potential crop trees in clearcut forests. I also suggest that if researchers and certifiers are to differentiate between avoidable and unavoidable damage, they
may need additional training to develop or at least to recognize technical

competence. Clearly, more research is needed on the effects of training in directional felling techniques and skidder operation, but insights are also needed on cost-effective incentive schemes for better logging. Where sawyers and skidder operators are paid on the basis of timber volumes yarded to the roadside, how can they be induced to treat the forest carefully? In many countries, command and control approaches to this problem seem particularly prone to being corrupted or at least not being implemented, but incentive systems are not well developed either.

Economic analyses of different approaches to forest management continue to be a priority, especially in the tropics where even basic financial data on harvesting costs are often not collected or made available. Environmentally concerned foresters are quick to refer to studies from Surinam (Hendrison 1990) and Sarawak (Marn and Jonkers 1982) that demonstrated the cost effectiveness of well-planned ground-based timber yarding. In these cases, due to the cost savings related to increased yarding efficiency in mapped stands with preplanned skid trails, well-managed harvesting operations cost less per cubic meter yarded to the roadside than uncontrolled conventional logging. These types of studies need to be repeated in other forests with more difficult terrain and contrasting socioeconomic conditions, where labor costs are higher, and at larger spatial scales before the generalizability of the conclusions are likely to be accepted by the logging industry. For example, where ground-based yarding is implemented on steep lands, there are likely to be lost revenues associated with foregone timber in areas deemed inaccessible by the harvesting guidelines. Also, wet-weather shutdowns of log yarding may be costly in terms of timber outturn rates where operations cannot be rapidly shifted to drier areas.

In conducting investigations on the costs of different harvesting methods, researchers need to clearly differentiate financial and economic costs and benefits. For example, by not blading down to dry soil on every bulldozer pass over a skid trail, yarding efficiency may be reduced (a financial cost paid by the logger). But, on the other hand, by facilitating skid trail closure operations, reducing soil erosion, and increasing post-logging forest recovery rates, this same restriction results in substantial long-term economic benefits to forest owners and society in general. A major challenge facing researchers in particular and conservationists in general is setting acceptable financial values for these and other noncommodity products and services. Some economists advise avoiding this neoclassical economic problem by not discounting the future and otherwise by adopting more environmentally friendly evaluation procedures (Gowdy and O'Hara 1995). Such novel approaches to economics should be investigated, but while the people who determine the fates of most forests still make their decisions based on traditional values of internal rates of return and net present values, we must continue to learn more about the financial costs of good forestry.

Silvicultural Effects on Commercial Species

Given the tremendous diversity of forest species, species-specific regeneration requirements, and stand histories and habitat types worldwide, it should be no surprise that silvicultural knowledge lags behind our understanding of forest engineering principles. Therefore, it seems quite reasonable to accept the *FAO Model Code of Forest Harvesting Practice* (Dykstra and Heinrich 1995) as applying to forests in general. In contrast, no such code is possible or desirable for silviculture. In fact, "legislated silviculture" (the application of a single silvicultural technique over large, politically defined areas) has been the anathema of good forest management. Government-mandated restrictions on conifer volumes harvested from mixed hardwood forests in Mexico, for example, substantially reduced recruitment of the light-demanding pines (Snook and Negreros 1986). Instead of a single silvicultural code for all forests, the challenge is to define general silvicultural principles from which trained foresters can prescribe treatments at the stand level. These prescriptions must be appropriate on the basis of site-specific management objectives, species' requirements, environmental conditions, logistical and technical constraints, and governmental regulations. Certifiers and drafters of certification guidelines must be knowledgeable about silvicultural principles and be able to recognize appropriate approaches to forest management. They must also be able to devise efficient and accurate methods for auditing management practices in specific field situations.

It seems worthwhile to step back and look at the range of silvicultural treatments available to foresters so as to better identify issues that need to be addressed by researchers interested in forest certification (Table 13.1). Basic silvicultural approaches can be grouped into those used in even-aged management and those used in uneven-aged management. Silviculturalists can also prescribe various post-harvesting treatments to increase stocking and growth of particular species or groups of species that share environmental requirements. A major challenge for even highly trained forest managers is in selecting combinations of potential treatments that are likely to prove financially profitable, silviculturally successful, and environmentally acceptable. Harvesting constitutes the most intensive of silvicultural treatments, but whether the effects are silviculturally beneficial or deleterious depends on whether the harvesting regime selected is appropriate and is applied in an environmentally conscientious manner. Single tree selection, for example, is ap-

Table 13.1

Basic silvicultural methods with emphasis on treatments likely to be of concern to forest product certifiers. Note that various combinations of these treatments are often prescribed.

I. Harvesting Methods
  A. Uneven-aged
    1. Single tree selection means scattered individual trees are removed
    2. Group selection means small clusters of trees are removed
  B. Even-aged
    1. Shelterwood is staged removal of entire stands over a small fraction of the rotation
    2. Seed tree is clearcut except that a few seed trees are retained
    3. Clearcut is removal of entire stands
II. Stand Improvement Treatments means enhancement of growth and value of potential crop trees or other commercial species
  A. Thinning
    1. Removal by cutting, girdling, or herbiciding understory means thinning from below
    2. Liberation thinning is removal of trees and vines competing with potential crop trees
    3. High thinning is removal of all noncommercial canopy trees in the canopy
  B. Weeding is removal of undesirable vegetation
  C. Pruning is mechanical removal of lower branches from potential crop trees
III. Seed Bed Preparation/Site Preparation
  A. Controlled burns to remove logging debris or to expose mineral soil
  B. Mechanical site treatments (such as roller chopping or disking)
  C. Herbicide treatment of vegetation interfering with seedling establishment or growth
IV. Artificial Regeneration Methods
  A. Direct seeding of native species in felling gaps
  B. Enrichment planting with native species (nursery-grown seedlings or wildlings) in felling gaps or along cleared lines through degraded or otherwise poorly stocked forest
  C. Plantation conversion
    1. Natural regeneration of commercially valuable species allowed to persist among planted trees (usually long rotations for timber)
    2. Only planted tree species favored
      a. timber trees planted after minimal site preparation and with infrequent stand tending over long rotations
      b. short rotations with intensive management

propriate for promoting the growth and advanced regeneration of shade tolerant species. Where properly managed, single tree selection disrupts few ecosystem processes (such as nutrient cycling and hydrology) and has only minor impacts on overall biodiversity. In contrast, where light-demanding species are targeted for management (e.g., *Swietenia macrophylla*), single tree selection is not readily distinguished from timber "mining" or "high-grading" (Snook 1994).

Stand "improvement" treatments (such as thinning) and site preparation operations (such as bedding) vary in intensities, frequencies, and spatial scales and consequently also vary in their on- and off-site effects. It would be helpful to know the effects of these various silvicultural treatments. Figure 13.1 represents a graphical attempt at displaying the range of effects of a range of forest treatments on biodiversity and ecosystem processes such as nutrient retention and carbon storage. A similar and probably equally contentious figure depicting the effects of agricultural intensification was presented by Vandermeer and Perfecto (1995); I am hopeful that Figure 13.1 will also provoke thought and discussion.

Predicting the effects of silvicultural treatments on nontarget species and ecosystem processes is particularly difficult because stand-level responses de-

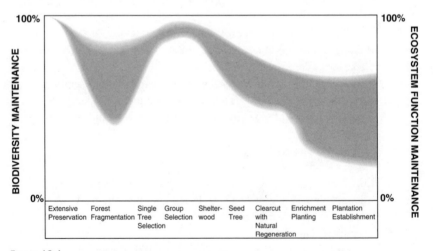

Figure 13.1
Hypothetical stand-level effects of forest fragmentation and silvicultural treatments of natural forest on biodiversity and some ecosystem functions (such as water retention, nutrient cycling, and carbon storage). The upper boundary reflects appropriate and carefully applied silvicultural treatments. Exceptions to the proposed trends are likely (for example, increased net primary productivity in intensively managed stands) but hopefully will be informative as well.

pend on more than just stand-level treatments; conditions in adjacent stands and over the entire landscape variously influence what happens in treated areas. Our knowledge is so limited that there is no forest in the world for which the total on- and off-site consequences of forest management are known. Based on experience, however, some of the consequences can be generally judged benign while others are likely to be deleterious. For example, where silvicultural treatments mimic natural forest dynamics, treatment effects on diversity as well as on hydrology, nutrient cycling, and other ecosystem processes are likely to be minimal (upper portion of band on Figure 13.1). Inappropriate or poorly implemented management practices, in contrast, can be very disruptive of ecosystem processes and substantially reduce biodiversity. Perhaps certifiers should focus on whether the chosen silvicultural treatments are applied with adequate environmental protection instead of trying to dictate which treatments should be administered. The pendulum has swung too many times between advocates of even and uneven-aged silviculture to be comfortable with politically motivated silvicultural mandates.

During the process of designing guidelines for certifying forestry operations or in carrying out certification audits, it is extremely disconcerting to realize how little is known about forest management, particularly in the tropics. Enough is known about silvicultural treatment effects, however, to suggest a research strategy that will provide a firmer foundation for sound forest management and forest certification.

Although every forest and every species is different, several common bottlenecks are faced by managers working toward the goal of sustainability. Securing and maintaining ample regeneration of commercially valuable species, for example, is a challenge faced by all natural forest managers. In Southeast Asian Dipterocarp forests, for example, natural regeneration (seedlings, saplings, poles, and small trees) of many canopy tree species is often sufficiently dense prior to selective logging to assure stand regeneration *if logging damage is minimized* (Wyatt-Smith 1987). A focus on reducing the impacts of logging in such forests is clearly justified and should be maintained. Where the valuable canopy tree species are not well represented in the understory, such as in many Latin American and African forests, harvesting methods and post-logging silvicultural treatments designed to enhance regeneration deserve more attention from researchers. Unfortunately, where the target species are light-demanding the range of possible management techniques available may be constrained, sometimes unnecessarily so, by environmental concerns. In particular, forest managers seeking certification may be restricted in their use of fire, pesticides, and clearcutting, all of which have potentially positive roles to play in some stands. Unfortunately, these same treatments have distinctly deleterious impacts under other circumstances, and deservedly bad reputations among many environmentalists.

Evidence is accumulating that there are few truly pristine forests in the world (Sprugel 1991). Many of the most majestic forests are now known to have regenerated after cataclysmic (stand-destroying) natural disturbances. Many more forests have suffered and recovered from anthropogenic disturbances during previous decades and centuries (Bush and Colinvaux 1994). If foresters are to maintain the species composition of managed stands in such forests and, for economic reasons, are to regenerate commercially valuable light-demanding species, they will probably have to mimic these cataclysmic disturbances (such as hurricanes and intense fires). Research on the minimum disturbances needed to regenerate canopy species is therefore of the highest priority in many forests of this sort. Since disturbances differ in intensity, size, and frequency, so should the experimental treatments. Also, given that fire is likely the most widespread disturbance affecting the regeneration and maintenance of many natural forests that are now under management, controlled burns may need to be assessed as a silvicultural treatment. In areas where frequent anthropogenic fires have resulted in forest degradation, prescribed burns might seem out of place but nonetheless deserve attention from researchers. Where fire was naturally frequent, as in many forests in North America, fire *prevention* can be viewed appropriately as a major disturbance. To the extent that clearcuts mimic former stand-regenerating disturbances (natural or otherwise), they too need to be investigated as silvicultural treatments for securing regeneration of light-demanding species. There is no need, however, for the same mistakes regarding clearcutting to be made by tropical foresters as were made by their temperate and boreal counterparts. For example, if patch cuts or group selection achieve the silvicultural goal of enhancing regeneration, there may be no justification for clearcutting large tracts of forest (beyond the generation of windfall profits).

Effects of Silvicultural Treatments

Every silvicultural treatment has effects that reverberate through the entire ecosystem. Even if the goals of the treatment are realized (for example, enhanced regeneration and improved growth of commercially valuable species), the negative effects on species composition, biodiversity, and ecosystem functions may not be acceptable to forest product certifiers. The studies necessary to determine these effects should address changes at the population, forest, and landscape levels. As a start, Martini et al. (1994) suggested a protocol for predicting the susceptibility of tree species harvested for timber to local extirpation; their approach warrants testing and further development to include nontimber species. If population "viability" analysis is carried out, for example, investigators need to consider population structure and change under the range of stand treatments to be applied (e.g., Peters 1994). Furthermore, population projection techniques based on matrix models, such as described by Caswell (1989) and McDonald and Caswell (1993), need to be modified to reflect increased understanding of natural history (such as nonstationary transition probabilities).

Research on nontarget effects of stand improvement treatments is also needed in many forests. Silviculturalists often prescribe these treatments to increase timber volume increments and wood quality in well-stocked stands. Data on the efficacy of the various thinning, weeding, and pruning methods employed by forest managers is largely lacking, particularly in the tropics. For the purposes of certification, however, research should focus on the effects of these treatments on nontarget species and ecosystem processes. For example, if selective thinning around potential crop trees (i.e., "liberation thinning") is recommended on the basis of the minimal environmental effects, it would be useful for certifiers to have data on treatment costs and silvicultural benefits relative to the killing or removal of all noncommercial trees and vines.

Research is also needed for developing guidelines for pesticide use in managed forests, especially in the tropics where relatively little is known about the use and effects of new and reportedly environmentally benign formulations (Ahrens 1994). Although pesticides (herbicides, insecticides, and fungicides) have been and continue to be misused, they may have important roles to play in certifiable forests. For example, if control of invasive nonindigenous species is a criterion for certification, there may be no financially viable alternative to selective and careful herbicide use.

An important goal of forest certification is minimization of the deleterious effects or maximization of the beneficial effects of forest management on wildlife. Researchers can help to identify the species most likely to suffer from management activities and can test methods for mitigating these effects. The focus should be on mitigation because all silvicultural treatments hurt some species while benefiting others. Unfortunately, most wildlife ecologists have restricted themselves to demonstrating the deleterious effects of logging rather than, for example, suggesting how *management* impacts can be lessened (for a review see Frumhoff 1995). In contrast, some studies (Greenberg et al. 1994 on reptiles in Florida; Thompson 1993 on forest interior birds in the central United States) used very different research methods to elucidate some of the varying effects of forest management on very different sorts of organisms. Not surprisingly, they come to some very different conclusions that need to be evaluated critically by forest certifiers. Conflicts between wildlife and timber management are not solely biophysical in origin—forest management effects on wildlife in many areas cannot be assessed without due attention to problems related to hunting by logging crews and increased forest access by other hunters.

Research that will provide the data necessary for drawing more informed lines on Figure 13.1 will involve scientists from a wide range of disciplines. For example, hydrologists need to continue testing methods for reducing erosion rates in logged catchments and developing better predictive models. Ornithologists need to address such issues as differences in edge effects resulting from group selection and clearcutting. Mensurationists need to monitor stand responses so as to inform debates about potential trade-offs between treatment benefits (such as increased volume increments) and treatment costs (such as reduced populations of understory frugivores). The band on Figure 13.1 will always remain fuzzy, broad, and contentious due to the incredible complexity of the underlying processes and great variation in the care with which silvicultural treatments are applied.

The focus of this chapter is on timber management, but harvesting of nontimber forest products can also have substantial impacts on biodiversity, ecosystem processes, and future profits. Although management for timber and for nontimber forest products are often compatible, they are often assumed to be in conflict (Putz 1992). Current efforts to develop certification guidelines tailored for nontimber forest products should provide further motivation for research on these potential compatibilities. Even ecotourism might benefit from good forest management under some circumstances.

Research on Uniquely Social Issues

The biophysical research focused on thus far in this chapter obviously needs to be conducted in parallel with socio-economic studies. Silvicultural treatments that are not cost-effective from the perspective of the people determining the fates of forests are unlikely to be widely applied. Furthermore, selection of silvicultural treatments for study should reflect the pertinent social conditions affecting the stands to be managed.

Many of the principles upon which forest certification is based relate to the various roles of forests in human society. The initial challenge seems to be in identifying which stakeholders to consider in any particular forest situation: all of the people of the world for whom forests represent part of the global commons; species other than *Homo sapiens*; governments that claim forests within their jurisdiction; concessionaires with rights to harvest timber; future generations; various forest "owners"; or, perhaps most clearly, forest-dwelling and forest-dependent indigenous people. Colfer (1995) suggested that participation in and benefits from the forest product certification process should be determined on the basis of proximity, preexisting rights, culture-forest integration, dependency, power deficit, and indigenous knowledge. Once the stakeholders are identified, researchers could assist certifiers in developing the most appropriate mechanisms for reaching consensus on issues regarding forest management, forest product processing, marketing, and profit sharing.

Conclusions

Emergence of forest product certification as a market-based incentive for better forest management indicates public dissatisfaction with the manner in which many forests are being treated. To an as yet mostly undetermined degree, consumers of forest products desire to use their choice of certified products to change forest management practices. Apparently the existence of environmentally sound and socially sensitive governmental forestry policies in conjunction with timber industry-monitored "best management practices" have not inspired consumer confidence that their purchases of forest products are not contributing to forest destruction and impoverishment of forest-dependent people. One source of funds for research on issues related to forest certification, therefore, should be the buyers of certified products.

Transferring all of the costs of certification to environmentally conscious consumers seems somewhat unfair, however, considering the wide range of beneficiaries of better forest management. Also, given the substantial profits enjoyed by some forest industries, relying on "green" consumers as the sole source of funds for research on forest certification seems unreasonable. Clearly the most equitable solution to the funding problem is costsharing by all of the beneficiaries of better forestry.

Collaborative approaches to research related to forest certification will maximize credibility, equitability, and utility. All forest stakeholders should contribute to setting research agendas and to paying the costs of data collection, analysis, and dissemination. Forest industries should participate substantially insofar as they are among the likely long-term beneficiaries of better forest management. Since the fates of forests greatly influences the future prospects for human society, international development and conservation agencies can easily justify their contributions to forest certification efforts. With its mandate to promote forest conservation and alleviate rural poverty through research, the Center for International Forestry Research (CIFOR) is an obvious source of technical support for forest certification research efforts (see Prabhu 1994).

Given their extensive experience and legal responsibilities regarding forest management, governmental forest research organizations also have a great deal to contribute. In light of their often intimate knowledge of the forests on which they depend, local people also can contribute substantially to forest certification research. In addition to being more informed, collaborative research has the advantage of being more likely to be accepted by all parties concerned than research carried out by any of the forest stakeholders in isolation. This acceptance will be enhanced by the extent to which investigators integrate the knowledge and experience gained over many decades of formal forestry research.

In preparing this chapter I frequently found myself contemplating research projects that would perhaps more appropriately be carried out by production-oriented forest researchers rather than by researchers specifically concerned with forest product certification. The distinction, however, is not always clear. For example, both certifiers and forest managers will benefit from knowing more about the silvics of commercially valuable species, but the research from which such knowledge would derive may not be specifically related to certification. This ambiguity is to some extent due to my being interested both in maintaining the commercial value of forests so as to increase the likelihood that they will be conserved, and in assuring that certified forest management operations are in fact environmentally sound and socially acceptable.

In the context of certification, definitions of "sound forest management" will be established by the accreditation agencies (e.g., Forest Stewardship Council). For these definitions to evolve rationally, well-integrated research by multidisciplinary research teams will be needed. Certifiers and researchers need to remain flexible and responsive to local conditions. Hence the emphasis in this chapter on silvicultural principles rather than on specific research projects.

Certification guidelines need to be stringent enough to protect forest ecosystems and society while not being so stringent as to make voluntary participation in certification programs unlikely. Striking a balance will require contemplation of the central question: How much disturbance should be permitted in certified forests? As a natural scientist, I probably should step back from this politically charged question and focus on the researchable silvicultural and ecological issues that will provide the basis for informed decisions. I do so with confidence that as we learn more about forest functions and relationships between forests and human society, certification guidelines will be allowed to evolve.

References

Ahrens, W. A., ed. 1994. *Herbicide Handbook* (7th edition). Weed Science Society of America. Champaign, Illinois.

Boyle, T. J. B., and J. A. Sayer. 1995. Measuring, monitoring, and conserving biodiversity in managed tropical forests. *Commonwealth Forestry Review* 74: 20–25.

Bradley, A. H. 1993. Testing a central tire inflation system in western Canadian log-hauling conditions. Forest Engineering Research Institute of Canada, Technical Note 197, 11 pp.

Brown, A. G., and E. K. S. Nambiar. 1995. *Management of Soil, Water, and Nutrients in Tropical Plantation Forests.* ACIAR Canberra, Australia.

Bush, M. B., and P. A. Colinvaux. 1994. Tropical forest disturbance: paleoecological records from Darien, Panama. *Ecology* 75: 1761–1768.

Caswell, H. 1989. *Matrix Population Models.* Sinauer, Sunderland, Massachusetts.

Cedergren, J., J. Falck, A. Garcia, F. Goh, and M. Hagner. 1994. Reducing impact without reducing yield. ITTO *Tropical Forest Update* 4: 9–10.

Colfer, C. 1995. Who counts in sustainable forest management. CIFOR Working Paper 7.

Dykstra, D. P., and R. Heinrich. 1995. *FAO Model Code of Forest Harvesting Practice.* Rome: Food and Agriculture Organization of the United Nations.

Evans, J. 1984. *Plantation Forestry in the Tropics.* Clarendon Press, Oxford.

Froehlich, H. A., and D. H. McNabb. 1984. Minimizing soil compaction in Pacific Northwest forests. In E.L. Stone, ed. *Forest Soils and Treatment Impacts.* University of Tennessee, Knoxville.

Frumhoff, P. C. 1995. Conserving wildlife in forests managed for timber. *BioScience* 45: 456–464.

Gowdy, J. M., and S. O'Hara. 1995. *Economic Theory for Environmentalists.* St. Lucie Press, Delray Beach, Florida.

Greenberg, C. H., D. G. Neary, and L. D. Harris. 1994. Effects of high-intensity wildfire and silvicultural treatments on reptile communities in sand-pine scrub. *Conservation Biology* 8: 1047–1057.

Hendrison, J. 1990. *Damage-Controlled Logging in Managed Rain Forest in Suriname.* Agricultural University, Wageningen, The Netherlands.

Jonsson, T., and P. Lindgren. 1990. *Logging Technology for Tropical Forests—For or Against?* The Forest Operations Institute, Sweden.

Marn, H. M., and W. B. Jonkers. 1982. Logging damage in tropical high forest. In P.B. Srivastava, ed. *Tropical Forests, Sources of Energy through Optimization and Diversification.* Pertanian University, Malaysia.

Martini, A. M. Z., N. de A. Rose, and C. Uhl. 1994. An attempt to predict which Amazonian tree species may be threatened by logging activities. *Environmental Conservation* 21: 152–162.

McDonald, D. B., and H. Caswell. 1993. Matrix methods for avian demography. *Current Ornithology* 10: 139–185.

Peters, C. M. 1994. *Sustainable Harvest of Non-Timber Plant Resources in Tropical Moist Forest: An Ecological Primer.* World Wildlife Fund/Biodiversity Support Program, Washington, D.C.

Pinard, M. A., and F. E. Putz. 1996. Retaining forest biomass by reducing logging damage. *Biotropica* 28.

Pinard, M. A., F. E. Putz, J. Tay, and T. Sullivan. 1995. Creating timber harvest guidelines for a reduced-impact logging project in Malaysia. *Journal of Forestry* 93: 41–45.

Prabhu, R. 1994. Assessing criteria for sustainable forestry. *ITTO Tropical Forest Update* 4: 6–8.

Putz, F. E. 1992. Unnecessary rifts. *Conservation Biology* 6: 301–302.

Putz, F. E., and V. Viana. 1996. Biological challenges of certification of tropical timber. *Biotropica* 28.

Snook, L. C. 1994. Regeneracíon y crecimiento de la caoba (*Swietenia macrophylla* en las selvas naturales de Quintana Roo, Mexico. In L.K. Snook and A. Barrera de Jorgenson, eds. *Madera, Chichle, Caza y Milpa*. Instituto Nacional de Investigaciones Forestales y Agropecuarias, Mexico City.

Snook, L. C., and P. Negreros. 1986. Effects of Mexico's selective cutting system on pine regeneration and growth in a mixed pine-oak (*Pinus - Quercus*) forest. In *Current Topics in Forest Research: Emphasis on Contributions from Women Scientists*. USDA Forest Service General Technical Report SE-46.

Sprugel, D. G. 1991. Disturbance, equilibrium, and environmental variability: what is 'natural' vegetation in a changing environment? *Biological Conservation* 58: 1–18.

Thompson, F. R. III. 1993. Simulated responses of a forest-interior bird population to forest management options in central hardwood forests of the United States. *Conservation Biology* 7: 325–333.

Vandermeer, J., and I. Perfecto. 1995. *Breakfast of Biodiversity. The Truth About Rain Forest Destruction*. The Institute for Food and Development Policy, Oakland.

Veríssimo, A., P. Barreto, M. Mattos, R. Tarifa, and C. Uhl. 1992. Logging impacts and prospects for sustainable forest management in an old Amazonian frontier: the case of Paragominas. *Forest Ecology and Management* 55: 169–199.

Wyatt-Smith, J. 1987. Red meranti-keruing forest. Malayan Forest Records No. 23.

Part III

Stakeholder Perspectives

It is clearly impossible to catalog the universe of opinions and perspectives on the issue of certification of forest products and management systems. Nevertheless, it is essential to present a wide range of such views. Since certification is such a recent concept, there are relatively few empirical and quantitative data on which to base conclusions. These divergent perspectives may therefore provide insight into some of the unanswered questions about certification.

In compiling the following sets of comments we sought to assemble what we believed was a representative cross section of current views on certification. In some cases, because of space constraints and inevitable redundancies, the editors suggested reductions in the initial submissions.

It must be recognized that views and opinions of individuals as well as organizations and governments often change over time, especially concerning the rapidly evolving field of certification. Therefore, the following compilation must be taken as one snapshot in time—certainly neither exhaustive nor definitive. Additionally, the views of the authors do not necessarily express the opinions of the organizations with which the authors work.

Of particular interest to readers may be the ongoing consultancies commissioned by the ITTO in 1995, which are examining the implications and potentials for certification, and the progress of the ITTO in encouraging their member countries (both producers and consumers) to become signatories of the ITTO and its Objective 2000. In addition, however, many bilateral donor agencies, as well as multilateral organizations, are separately investigating the role that certification may play in their development strategies regarding forest management.

The interest and activity regarding certification are high. We hope this can at least serve as a catalyst for further constructive dialogue among all interested parties.

❦

Conservation NGOs

National Wildlife Federation

In the cranky, stubborn seasons of New England, we "get our wood out" when we can. The best time—if you don't mind the sleet, wind, or bone-cracking cold—is in the winter when the roads are frozen stiff and the logs, free of slash, can be skidded clean. But it is not unusual to be harvesting and handling wood year round, since mills are always adjusting inventory, the markets are often in flux, and there is always firewood to put up. Shake the hand of a Vermonter or a Mainer and you're bound to get splinters!

In the northeastern United States, our forest products economy is a mixture of commodity-driven forestry, family-owned lands and mills, and hundreds of private small-lot forest landowners. Publicly owned land across the four northern New England states is scarce, comprising less than 15 percent. From this mix flows a predictably uneven stream of management goals, environmental impacts, and harvesting techniques. Overall, the dominance of private landholders in our region and in this forest type is attractive to the goals of forest certification.

But how will certification fit in this regional rhythm, and how are environmentalists likely to respond? One thing is clear: *We need alternatives.* Our attempts to coax good forest practices through tax incentives, land-use planning, and tepid regulation have been marginal at best. Although sustainable forestry is being actively discussed—and occasionally practiced—in our region, much forest land has been impaired by a continuing legacy of high-grading, clearcutting, and carelessness.

Certification will appeal to the New England preference for voluntary, market-driven solutions to the extent we can work past inevitable hurdles, such as:

• *Noise in the markets.* Almost every major forest products company competes for the green consumer by pushing environmental messages or logos in their packaging, backed by Madison Avenue advertising. New labels from timber and other environmental product certification programs will add to the din.

• *Uncertain public-sector response.* Some government agencies may be threatened by independent certification, while others will embrace it. One bad actor in the field (for instance, environmental violations or a case of mislabeling) could bring an avalanche of government oversight under the guise of consumer protection—leading to differing regulations affecting certification in different places.

• *Fragmented participation.* If only small, low-volume firms participate in certification it will seem trivial or quaint, whereas if only big firms are involved it will seem unfair and invite suspicion about the quality of forest management.

• *Illusion of timber security.* Many environmentalists worry that certification will sanctify a spiraling global appetite for wood; they argue that current wood consumption must be limited if we want to protect necessary levels of biodiversity and other services of forest ecosystems.

Environmentalists are naturally suspicious about other things as well: that certification will be co-opted from the inside by industry; that only marginal improvements in forestry will be achieved where abrupt sea-changes are necessary; that market forces can waylay even the best forest management intentions; and that certifiers may have trouble policing themselves. It is easy to stand back and grouse about the risks—no doubt some of them will be real at some time in some setting.

But the risks to forest health and biodiversity from muddling along with our current practices are just as bad and probably worse. Certification deserves earnest if only grudging respect from the environmental community and a reasonable time to work. This is why the National Wildlife Federation is involved directly in the development of nonprofit, third-party certification in North America and specifically in the northeastern United States.

We believe timber certification has strong merit in helping define sustainable forestry and, by extension, locally sustainable economies. We hope certification will serve to recognize and induce a much higher standard of forest management than has been typically practiced on private lands and, in turn, promote a more ecosystem-based ethic of land ownership. We know that better mechanisms are needed to communicate the true costs (and savings) of good forestry in the marketplace. While no one management or marketing approach can meet all our aspirations for how forests are used, this concept—faithfully followed—can satisfy many of them.

🐦 **Eric Palola**
   Resource Economist, Northeast Natural Resource Center, NWF
   Montpelier, Vermont

Forest Activists

The increasing public demand for credibly sustainable sources of wood shows signs of being irreversible. The continuous denouncements of forest devastation worldwide (but mainly in the tropics) by environmental groups has raised the attention of the public and brought great pressure for a solution that would accommodate the preservation of forests and the consumer needs of industrial societies. One of the most pointed solutions to emerge from worldwide forest campaigns is the certification of forest products. However, there has not been enough discussion on the need to curb superfluous consumption of forest products. The search for sustainable sources of wood has to be encouraged but without losing the perspective that it is necessary to intensify the discussion on consumption.

The idea of the Forest Stewardship Council (FSC) is extremely challenging, trying to accommodate under the same umbrella answers to the technical questions regarding sustainable use of forests (ecological adequacy, social fairness, and economic viability), political strategies for encompassing all forest regions of the planet, and the establishment of a legitimate participation of different sectors of society.

It is inevitable that such a complex process would lead to locally different paces of development, depending on market pressure, local interest from the government and society, financial support, and access to information. Given the continental dimensions of Brazil, such a task could either be transformed into a set of immense problems, or it could lead to a significant demonstration of the feasibility of the approach. Some of the first steps of the FSC process in Brazil have been taken, but there is ahead a great deal of work that requires a well-controlled strategy.

The continuation of FSC activity in Brazil should start with a separation between the certification of products from natural forests and the eventual accreditation of monocultures of exotic tree plantations. Both sources of wood are extremely important in the context of the economy of Brazil, given the dimensions of Amazonia and the continuous expansion of fast-growing tree plantations for industrial wood production. However, in spite of the fact that FSC involvement in Brazil has so far been pushed by the interests of the pulp and paper industry, there are no principles and criteria for sustainability yet approved for tree plantations. Also, these already controversial exotic tree plantations raise the additional issues of product life cycle and the inevitable questions about pollution associated with the processing and use of this material.

For the FSC, finding certifiable sources of wood from natural forests such as in Amazonia, the principles and criteria will have to be implemented at the local level. My intuition, after almost 10 years of work in Amazonia, is that a missing key concept in the FSC Principles and Criteria will then appear— the landscape perspective. Natural regeneration of Amazon forests after interventions depends on the surroundings, which depend on creating alternatives to migratory (slash and burn) agriculture for local populations and on halting present destructive forest practices and the unplanned expansion of frontiers.

Finding adequate technical approaches to forest management in Amazonia is only the first necessary move. It will only become widespread if chain-of-custody can be clearly established for all forest products coming from the region. At present, the lack of control from governmental agencies, due to many reasons beyond this discussion, is causing extreme pressure on NGOs for filling the watchdog role in this process. As yet, there is no mechanism within the FSC for handling grievances and appeals. Although this issue may seem premature, in view of the recent commencement of accreditation of certifiers, there is a serious danger of the process being bulldozed by large economic interests that are already pushing for certification from well-known certifiers. Only the establishment of a full FSC office in Brazil, with local principles and criteria and a local governing board, will be able to appropriately tackle these challenges. Foreign initiatives will always involve the risk of not understanding the issues adequately and/or not being accepted by the different sectors of the society.

Projections for Amazonian participation in the international market for tropical woods are quite alarming, especially given that the resources from Asia and Malaysia are being quickly depleted. Domestic consumption of wood is also rising dramatically, meaning that simple demands for boycotts may have to substitute for serious discussion of nonwasteful use of forest resources from sustainable sources. The isolation of producer countries such as Brazil from consumer markets driven by concerns with forest preservation cannot be allowed to take place, due to the serious risk of the last remaining tropical forests being quickly devastated. The FSC, having started this debate, has the obligation to ensure equitable development of alternatives for sustainable use of forests in all parts of the planet.

❦ Anna Fanzeres
   Forest Engineer, Greenpeace
   Rio de Janeiro, Brazil

Global Forest Policy Project

The rapid emergence over the past few years of *independent third-party* forest certification programs, and of the Forest Stewardship Council (FSC), has changed the nature of the entire international forest policy debate. The change has affected not just timber trade discussions, but also a wide range of initiatives dealing with all aspects of forest policy, including the very definition of "sustainable forest management" and how to monitor and measure it—and even discussions about a possible new global forest treaty.

This change has occurred across the full spectrum of institutions and policy fora. For example, several intergovernmental processes have sprung up to create sets of "Criteria and Indicators" (C&I) for assessing, at the national level, the sustainability of a country's forest management. Although it's too early to tell whether these initiatives will benefit the world's forests, many environmental groups believe some of the governments involved would like to use such C&Is as a substitute for, or even as a shield against, independent certification.

Another initiative has involved the International Organization for Standardization, which in June 1995 debated for the first time a proposal to develop a set of "sustainable forest management standards." Environmental groups, who saw this as an attempt by some in the timber industry to undermine independent certification programs, helped block the proposal.

Indeed, every multilateral forest policy discussion since the 1992 Earth Summit in Rio de Janeiro has addressed the certification issue in one way or another, and the most prominent new initiative, the UN Commission on Sustainable Development's Intergovernmental Panel on Forests, features certification high on its agenda.

At the national level, in countries all over the world, independent certification has generated a great deal of interest among governments and the timber industry. Some support it and some don't. In the United States, a timber industry association has launched a sustainable forestry, self-assessment initiative among its member companies. Some environmental groups believe the initiative is a direct response to the rise of independent certification programs, which the association opposes.

What is it about independent certification that has generated such attention?

First, the concept of *independence* from government and industry makes some in government and industry uncomfortable. However, it is exactly such independence that is essential to ensure the credibility of sustainability claims. So-called self-certification (a claim made by someone with a direct financial or political interest in the use of the claim) tends to have very low credibility with the public.

Second, credible certification is *performance*-based and *management-unit* focused. Members of the conservation-minded public want to know that the products they buy come from forests that (1) are still there and (2) have been proven to be sustainable, or well managed. This requires that there be measurable evidence of concrete results on the ground, in a specific forest, rather than vague or sweeping assertions of "changes" or "progress" being made, of "forest cover" being maintained, of management "plans" or "systems" being in place, or of forests being "rehabilitated" or "replanted."

Third, and most important, *the Forest Stewardship Council has emerged* to coalesce the general concept of third-party certification into a powerful, global political force. By pulling together several different components, political constituencies, and certification initiatives into a single system, the members and supporters of the FSC have become influential enough to begin to make, change, and, when necessary, prevent forest policy at the international level. Indeed, from my vantage point in the middle of the global forest policy debate, *the FSC is responsible for putting independent certification on the political map* and for taking the lead in drawing new lines on that map. Some in the forest products industry trade, however, would prefer that certification were not on the map at all.

In practical terms, it can be said that the FSC is writing the world's definition of sustainable forest management. While it's true that other organizations are writing their own definitions as well, the FSC is clearly setting the pace at the international level. It seems that everyone else is rushing to catch up (although at the national and local levels several exemplary models already exist). No defining process has been nearly as extensive, as diverse, and as inclusive in its consultations and its intended application as the FSC—nothing even comes close. And in my opinion none has come as close to describing the elements, the bounds, and the processes of environmentally and socially sustainable forest management as the FSC has.

In spite of its leadership in this field, I do not believe the FSC's current (1995) guidelines can guarantee absolute sustainability, although they will surely have exactly that result in many cases. However, I cannot see at this time any other international process that seems as likely to pull the world closer to forest sustainability, as rapidly, as the FSC is potentially capable of doing. And I am absolutely convinced that few other entities have changed the political debate over the world's forests—for the better—as much as has the FSC.

The impact of the FSC has not only been at the international policy level. The FSC has catalyzed, in an increasing number of countries around the world, national-level dialogues among diverse "forest stakeholders" where no dialogue, or only poor dialogue, existed before. And because the country level is where forest management decision making has the most impact, this alone is a significant achievement. The creation of the FSC has enabled environmentalists and forest and wood users to come together not just for dialogue, however, but to create a common mission and a common strategy for accomplishing it. At the same time, because the FSC has drawn so much attention from governments and industry, those stakeholders, too, have been pulled into the same dialogue.

I would like to emphasize that the FSC and independent certification are designed specifically to improve the management of forests that produce marketable products. They cannot address all issues of forest management or solve all the world's forest problems. For example, certification is not a substitute for national parks or other protected areas. However, I sincerely believe that the ultimate influence of independent certification will prove to be far wider than the certified forests themselves. As sustainable forest management is further defined by the FSC and others, and new incentives are provided, improvements in management could well spread to forests all over the world that are managed for a wide range of other purposes.

It also is important to emphasize that certification should not be seen as a restriction; it's an *incentive* to achieve sustainability. The FSC is based on an entirely voluntary approach to certification. If forest managers want to be rewarded for good management, they can ask to be certified; if not, they are free to manage their forests any way they wish. Furthermore, if as we all hope and are working for, certification can make sustainable forest management financially attractive, then forest managers will clearly have an incentive to earn certification. Through certification, they may be able to increase their profits, increase market share, and improve their public image. To forest managers, this ought to seem far preferable to the threat of boycotts, lawsuits, or new legal restrictions, which are unlikely to provide any of those benefits.

If solving the world's chronic, seemingly intractable forest problems were easy, we might have done so by now. Everyone is looking for solutions. Concerned citizens who are sincerely frustrated at witnessing the spread of global deforestation are willing to use almost any tool they can find to stop the destruction. Available tools include everything from filing lawsuits and enacting new laws, regulations, taxation, or tariffs to staging demonstrations and calling for timber bans and boycotts. *If we cannot suggest any alternative that will produce concrete results, we might as well get accustomed to the other tools and methods these concerned citizens may use.*

Independent certification may be such an alternative. It's an alternative that doesn't turn the world upside down or "the system" inside-out or create any "new world order." In fact, it can be said to fit quite easily into the existing socioeconomic market system, where rules are well known and widely accepted, where the "consumer is king"—which consumer, with enough information, is able to freely choose to "buy or pass by" any product he or she wishes.

If independent certification fails, the choice of remaining tools will seem to be a poor one for a lot of desperately concerned citizens, who may begin to see more radical options as increasingly rational. From the viewpoint of those in the forest products industry, the choice these citizens end up making may not be to their liking. By comparison, independent certification may seem altogether preferable. We're at a turning point.

❦ William E. Mankin, Director
    Global Forest Policy Project
    Washington, D.C.

Good Wood Alliance

When it comes to buying wood or specifying wood products for construction, most woodworkers, architects, and consumers consider three things: (1) How does it work? (2) How does it look, feel, and smell? And (3) how much does it cost? Lately, a fourth issue has been added to the equation: Where does the wood come from and how well is the forest managed in which it is grown?

This development is tentative but significant in that it signals the introduction of environmental responsibility into a commodity market that has functioned for centuries in the near-total absence of sustainable behavior. Historically, whenever timber supplies were depleted in one area, new forests were simply exploited elsewhere. In the United States, one can chart the westward migration of populations in the 19th and 20th centuries alongside the devastation of forests in New England and the Midwest.

That doesn't mean, however, that the emerging interest in well-managed wood products is incompatible with market forces. Indeed, it bears an important relationship to one very traditional component of the timber trade: quality. Woodworkers have long staked their reputations and their livelihoods on the quality of their material and the reliability of their source.

The extension of quality from products and materials to environmental performance makes sense, given the current pressure on natural resources. But the emerging "green market" faces a yawning chasm of information. Ethical and ecological aspects of forest management and wood procurement are much more difficult for wood users to judge than the tangible attributes of performance, appearance, and price mentioned above. This is especially true when wood is harvested and traded widely all over the globe, as it has been for centuries. The problem is exacerbated by a fundamental lack of transparency in the trade.

Independent forest certification can help. Only when trade operates with a higher level of transparency can consumers and manufacturers assume greater responsibility for the environmental impact of their wood consumption. Producers hold an obvious vested interest in promoting their products and, in most parts of the world, they have a dismal reputation for responsible forest management. The new industry of forest certifiers that has emerged to independently verify management claims can effectively bridge that credibility gap. Without certification, it is virtually impossible for the average lumber retailer or wholesaler (not to mention the consumer) to evaluate the management program of their suppliers.

The problem doesn't end with independent certification. Certification is, by definition, site specific, and the wood products industry is so large and so diverse that many different certifiers will be required to evaluate management practices around the world. What's more, certification must be implemented locally, in close collaboration with area residents and institutions, if it is to gain widespread support and credibility. To cope with this likely proliferation of certifiers, delegates to WARP I (The Founding Conference of the Woodworkers Alliance for Rainforest Protection, now known as the Good Wood Alliance) first identified the need to establish consistent criteria that all certifiers would be expected to uphold.

The organization that emerged from that series of urgent hallway huddles at the WARP conference, which took place in 1990, is today known as the

Forest Stewardship Council (FSC). Although the FSC represents a broad, international coalition of environmental organizations, private and public foresters, indigenous peoples' groups, wood producers and manufacturers, its roots are based firmly in the market. It was founded in response to a perceived need to connect quality, transparency, and responsibility at every level of the timber trade.

Admittedly, this is an untested formula. It is by no means certain that the market will reciprocate and that good forest management will flourish—even if we are successful in bringing quality and transparency to the harvest, manufacture, and sale of wood products. The only certainty is that forests and trees and the lives of those they support will be destroyed in ever-increasing numbers if our best efforts fail.

❦ Scott Landis
   Good Wood Alliance
   York Harbor, Maine

❦

Forestry Professionals

International Union of Societies of Foresters

Two factors played a key role in identifying and resolving international concern for sustainable forest management. First, under the auspices of the International Tropical Timber Organization (ITTO), a general commitment was made by the producing countries that all international trade of tropical wood products would come from sustainably managed forest units by the year 2000. The countries of the temperate and boreal regions opposed extending the year 2000 goal to their forests during the 1993 International Tropical Timber Agreement negotiations. However, in June 1993 at a meeting of the Helsinki forestry group, the United States officially committed to the sustainable management goal by the year 2000.

Nevertheless, the parameters of sustainable forest management as historically defined are now being widely questioned. Thus a commitment had been made to an as-yet-to-be-defined strategy.

Second, the development of the Forest Principles and Agenda 21—Combatting Deforestation (ch. 11)—under the UNCED contributed to formulating the elements of sustainability. However, both documents comprised only general guidelines developed for global application, thus leaving a need for refinement at national levels.

The development of suitable criteria and indicators for sustainable forest management raises the inevitable issue of verification of the results by independent third parties, or certification organizations. However, numerous standards have been developed that reflect on the credibility of that judgment.

The need for the inclusion of forestry professionals in sustainable development demands their involvement in the development of national criteria and indicators for sustainable forest management. Forests as a vital subset of a country's national environmental action plan can no longer be ignored as country strategies are produced.

Political and economic decisions have caused forest lands to be used for too long as a "social value" to move landless farmers into frontier areas without addressing land reform and property rights problems. As a result, many foresters are employed in situations where professional judgments are restricted and where their advice concerning sustainable management has usually not been acknowledged or followed.

A framework of criteria and indicators that utilizes ecosystem productivity and health under adaptive management would allow science, experience, and judgment to establish social, economic, and ecological sustainability. This process establishes guidelines that can judge sustainability for all types of forests—natural, secondary, plantations, or deforested areas—as each land management unit is integrated into the landscape level. Sustainability is then based upon each type having its own established set of criteria and indicators. These standards would be established at the local level, thus allowing the forestry professional to judge whether or not a particular management system was sustainable.

Under the guidelines of such management, conversion (harvest) would be permitted under specific circumstances and intensity from natural, semi-natural, and plantation forests. Logically, reforestation would be required under the forest management plan. There may be restrictions on rapid and large-scale cutting of any particular forest type. Within these guidelines the local community would affirm the action of the forestry professionals as they judge whether a particular management plan can meet certification standards.

Forestry professionals need to support forest principles and their application in partnership with local communities. Foresters have an ethical responsibility to bring to the attention of the local, national, and international community those cases where the principles are not followed.

The "command and control" mode of working with local communities is not compatible with modern concepts of local participation, especially with respect to equal representation by women. Forestry professionals need to challenge these structures. A new role for forestry professionals as facilitators and mediators in the equitable distribution of natural resources should be encouraged.

Certification is an important link in a variety of endeavors to find long-term solutions to better forest management. In regions where forests are mainly used for local fuelwood consumption, the effects of certification could be limited. But when viewed from a worldwide perspective, certification of forest management is part of a vital effort that needs the expertise of the forestry professional. All stakeholders need to join in a holistic approach to natural resource use where justice and equality are as vital to the solution as technical expertise.

❦ Howard Heiner
   Washington, D.C.

The Society of American Foresters (SAF)

In 1994 the SAF formed a Forest Management Certification Task Force to report on the implications of national and international certification. The conclusions of their report are listed below and reflect the attitudes toward the certification process of a cross section of foresters in the United States.

1. The discussion of forest management and forest products certification programs is increasing and will continue. As current efforts toward sustainable forestry continue to evolve, the role of certification should be further explored and defined.

2. Certified wood products are not a large segment of the wood market; however, public awareness and demand could continue to provide impetus for the certification movement. The necessity and appropriateness of certification programs in the effort to sustainably manage forests will ultimately be determined by the market place and public demand.

3. International and national governmental and nongovernmental discussions on standards and indicators of sustainable forest management will continue and will significantly shape the outcome of the forest management certification debate. As these dialogues to determine appropriate indicators and measures for sustainable forestry continue, the U.S. forestry community must actively participate to ensure that they are compatible with the unique forests, socioeconomic systems, and existing laws of the United States.

4. While there is value in agreed-upon underlying principles, standards, and indicators, any certification or quality assurance system must be flexible enough

to adapt to different and changing ecological, economic, and sociopolitical situations.

5. Certification is a mechanism through which some aspects of "sustainability" can be monitored. In turn, a market demand for sustainable products may create a market incentive for individual companies to certify products in order to increase competitiveness.

6. Certification programs may provide consumers with the ability to generate an incentive for the forest industry to practice sustainable forest management, although the willingness of consumers to do so is somewhat suspect.

7. Certification programs can provide a vehicle for industry to communicate positive accomplishments to the public.

8. The plethora of existing and emerging forest management certification organizations causes confusion, not only among final consumers of forest products but also with the producers of forest products.

9. Despite current problems regarding forestry certification programs, they are one of many potentially viable mechanisms to aid in promoting sustainable forestry in the United States and abroad.

10. Forest management certification is not likely to replace existing appropriate forestry regulation, but it could help shape or reduce the impetus for additional regulation.

❧ Society of American Foresters 1995
SAF Forest Management Certification Report, 14 pp.

An Industrial Forestry
Perspective on Forest Certification

The question of independent certification of forest practices has generated much discussion among U.S. forest landowners and forest industry managers. Given the historical evolution of our forest practices, and the many public and private sector institutions supporting sustainable forestry in our country, most forest landowners question why certification should be considered in the United States.

Much of the frustration with the idea of independent certification stems from the apparent misperception in Europe that forests in the United States are unregulated and presumably not being managed on a sustainable basis. Our forest landowners and managers have a much different perspective, based both on an understanding of recent developments in U.S. forestry and on the reality of managing to meet the American public's expectations.

To understand this point of view, it is essential to know that 59 percent of the commercial forest land in the United States is owned and managed by individual nonindustrial private landowners. Only 14 percent is owned by industrial companies, with the balance, 27 percent, in public ownership, principally in our national forests.

It is also necessary to understand that, while some federal laws affect private forestry, regulation of forest practices in the United States is primarily the responsibility of state government. Some states, including California, Oregon, and Washington, have developed comprehensive forest practices laws and regulations to address all aspects of private forest management. Other states rely on voluntary "best management practices," developed by forest landowners and state forestry agencies to protect water quality, riparian habitats, and other features deemed important in individual states.

In addition to state and federal requirements, the member companies of the American Forest and Paper Association have developed their own statement of Sustainable Forestry Principles, which they are pledged to implement on their own industrial forest lands and to promote among nonindustrial private landowners and wood suppliers.

Given the processes that are in place to promote sustainable forestry in the United States, most members of our industry question whether the incremental improvements in forest practices that might be gained through certification are worth the added expense, duplication of effort, and management complexity that any certification system would require. Few producers believe that the added expense of certification will be rewarded in the marketplace with a price premium for "green" products.

The requirement for chain-of-custody tracking is particularly objectionable, given the large number of private landowners who supply our industry and the multiple sources of raw material that most manufacturing facilities depend on. Furthermore, forest management decisions are the responsibility of the individual forest landowner, not forest products manufacturers who purchase stumpage or logs to supply their mills.

We know from experience that individual private landowners respond much more positively to education, technical assistance, and financial incentives than they do to external pressure or regulation. An excellent example of a successful private forestry assistance program is the American Tree Farm System, established by the forest industry in 1941, which now includes over 72,000 landowner-members, managing 35.6 million hectares of private forest land. Similar forest stewardship programs are sponsored by our federal and state governments. We believe that these landowner-oriented programs are a more effective means of improving forest practices on private lands than would be coercive certification programs requiring chain-of-custody tracking.

The U.S. forest industry is also quite concerned that independent efforts to promote certification will result in multiple sets of principles and criteria for sustainable forestry, causing confusion among customers, consumers, producers, and forest landowners. For this reason, the U.S. forest industry has supported our government's efforts to develop internationally acceptable criteria and indicators through the Montreal Process, most recently expressed in the February 3, 1995, "Santiago Declaration." If there are to be international principles and criteria for sustainable forestry, we believe that they must have government-to-government sanction and support in order to maintain an essential harmonized international trading system for forest products.

One other internationally recognized process, provided through the International Organization for Standardization including the new ISO 14001 Environmental Management Systems process, also shows promise as a method by which forest products companies can demonstrate their commitment to sustainable forestry. For the companies in the forest industry that are ISO 9000-certified, use of ISO-based processes will likely be a preferred means of demonstrating their progress in implementing sustainable forestry.

In summary, while the members of the forest industry in the United States remain committed to advancing the practice of sustainable forestry, they will remain wary of certification proposals that add cost and duplicate existing efforts, or that fail to gain acceptance among nonindustrial forest landowners. Any system of principles and criteria for sustainable forestry must be developed in a manner that is recognized and supported by forest landowners, manufacturers, and government and must maintain an essential level playing field for international forest products trade.

❧ John P. McMahon, Vice President
   Timberlands, External and Regulatory Affairs
   Weyerhaeuser Company, USA

❦

Organizations of Indigenous Peoples

Menominee Tribal Enterprises

The Menominee have an incredible strength in their integrated forest management and processing facility. Management of the Menominee Forest with sustained yield practices has presented challenges as well as advantages both in marketing and in processing. Processing and marketing are always tied to the resource and the annual allowable cut. While this can present some interesting challenges for marketers, it has positive ecological consequences for the forest, because market demand does not affect management decisions, but rather marketing responds to the resource. Basically, the forest drives the mill—the mill does not drive the forest. Management of the forest is geared toward growing quality timber while preserving the diversity of the forest.

Managing forests for timber alone under-utilizes potential income and benefits. Nontimber forest products, such as medicines, food, or crafts, are not providing tribal members with much income at this point, and it is uncertain whether individual members are interested in pursuing this opportunity. Tribal ventures such as tourism, which supplements the White River Apache's Forestry operation, have not been explored.

The forest enterprise is considered by the Menominee to be the backbone of their economy, and there is general community consensus that the forest needs to be maintained for economic reasons. Yet, it is probably the Menominee dependence on the forest for cultural survival that has provided the real motivation to maintain their forest. If this is true, and the forest survives in its present condition because of traditional cultural beliefs, any cultural devaluation is a potential threat to the forest.

The forest has provided the Menominee people with employment, economic benefit, and a strong tie to their cultural heritage. If we look at the reservation as a whole, including humans and their needs as part of the ecosystem, forest harvesting has not been detrimental. Current management practices using sustained yield forestry do not seem to compromise ecological sustainability. At this point, without more information, we can only look to the forest and see a healthy productive system and assume that our history proves that what we are doing works. For over 100 years, forest management has maintained both the forest and the people, and we expect that it will continue to do so.

(Adapted from *Forestry in the Americas: Community-Based Management and Sustainability*. Institute for Environmental Studies and Land Tenure Center, University of Wisconsin).

❦ Paula Rogers Huff and Marshall Pecore
   Neopit, Wisconsin

An Alternative to Logging
in the Pacific Islands (Soltrust)

"Ecoforestry" is a viable alternative to logging—environmentally, socially, and economically. It is an ethical and interconnected system of development that addresses the social, ecological, and economic requirements of people for sustained forest resource utilization and management. It challenges the current over-exploitation of human and natural resources. It aims at preventing environmental destruction, social injustices, and corruption that badly affect traditional ways of living; in particular, it aims at stopping forest destruction and degradation.

It is Soltrust's firm conviction that the present concept of "development" in the Solomon Islands, as far as forest resources are concerned, is inappropriate. It is accelerating the demise of forest resources that would be exhausted within a short period if it is not stopped or slowed down now. It is time to adopt a more responsible, moral, ethical, and rational approach to natural and human resource utilization and management in the Solomon Islands.

Ecoforestry is a progressive and acceptable form of resource management. It is environmentally safe and economically sustainable both in theory and practice. The same approach can be adapted to other forms of development, like ecotourism. It aims to overcome and or avoid all negative aspects of large-scale development. Ecoforestry as proposed by Soltrust aims at radically altering the perspective and thinking of politicians, business persons, councilors, elders, community leaders and decision makers, students, and the general public.

The aims of the Soltrust Ecoforestry program are:

1. To educate landowners, the government, and NGOs concerning the need for and importance of appropriate small-scale community-based developments like portable sawmilling projects

2. To encourage ecoforestry practices among forest owners as a development alternative to large-scale logging

3. To design and implement effective training methods in all aspects of development of successful small-scale portable sawmilling projects at the village level

4. To support village/community projects by providing training, technical backup, and a marketing support structure

5. To substantiate and prove that this kind of development is a more acceptable, appropriate, and beneficial way of managing forest resources, particularly where customary landownership prevails

Soltrust and Certification

As an independent division of Soltrust, IUMI Tugetha Holdings Limited (ITHL) has been created to act as a watchdog, particularly for project models that will be involved in the timber trade. With its independence, this division has the authority to suspend any timber activities between ITHL and a project model if there is good evidence showing abuse or violation of the forest management and environmental standards that have been mutually imposed upon them.

Certification of timber is of great importance to overseas buyers such as the Ecological Trading Company, Ltd. in the United Kingdom and the Netherlands, B & Q in the United Kingdom, Espen in Germany, and Ecotimber International in the United States. These companies demand proper certification by Soltrust of the timbers exported to them, so that they can promote the product as an environmentally and socially friendly product from the Solomon Islands.

❦ Antony Carmel
    Executive Director of Soltrust, Solomon Islands

Businesses

The Home Depot

Certification by an independent third-party scientific organization continues to be the foundation upon which The Home Depot is successfully building its environmental franchise with the customer and through which it is aggressively pursuing the interrelated goals of promoting alternative merchandise and delivering environmental information. Since 1991, The Home Depot has used the competitive dynamics of the marketplace to position almost 30 voluntarily certified nonforest products in its stores. Many of these represent new "green" products and product improvements. Other products represent environmental achievements that, until certification, their manufacturers had no credible way of claiming as having differential advantages. We have achieved this track record in a country, the United States (we also operate in Canada), that is not among the 25 or so countries with a government-sponsored national eco-labeling scheme. Instead, we have relied on the private sector to help create a de facto national eco-labeling scheme. As a result, more certified products can be found in the $140-billion U.S. hardware industry alone than in some national eco-labeling programs.

Certification has helped manufacturers maintain, enhance, and in many cases restore the credibility of environmental marketing claims and differen-

tiate their products based on environmental attributes. Most important, it has
helped empower customers to make more informed choices at a critical time:
Over the past four years decisions based on a product's environmental benefits is the only area of consumer decision making that has shown growth. As
lifecycle analysis becomes a more accepted and understood tool, it has even
appeared in our stores on product labels in the form of a full-disclosure certified "ecoprofile."

Another key to The Home Depot's strategy has been to avoid customer
confusion by specifying one eco-label, since in the United States there are
two competing groups. With reciprocity among FSC-accredited certifiers,
we are able to reinforce this strategy of having one "eco-brand" stand for
something in our stores.

It took us a while to gain momentum in certifying nonwood products. We
are now gathering that same momentum for forest products. However, there
are fundamental differences between the eco-labeling of nonwood (nonforest) products and forest products that have the potential to derail the progress
we've made. For nonwood products, eco-labeling takes the form of a single-
attribute claim or a participatory minimum standard set in part by those it
governs. But for forest products, the FSC-accredited certification system represents true third-party oversight.

True independent third-party oversight of the environmental (versus financial) "worth" of an asset like a forest is extremely threatening to many producers. It can be threatening even to governments, which will always act in
the interest of their populations and can regulate so that short-term social
needs, such as cheap wood for domestic housing or international trade and
development, outweighs the need for sound environmental oversight of their
forests.

Unfortunately, certification of forest products has been much more difficult
than certification for nonwood products. In a wood-scarce and price-con-
strained economy, one in which wood alternatives have not yet gained widespread acceptance, large U.S. retailers do not have as much maneuvering
room to change suppliers, particularly their North American ones. And the
highly leveraged forest products industry has become trapped, having to rely
on its forests to meet financial performance requirements first and environmental ones second. Certification could change this paradigm, so naturally it
presents a potentially threatening scenario.

Thus, it is natural that despite pressure from retailers to "do the right
thing," and through certification enable retailers to tell their suppliers' won-
derful stories of stewardship in a credible manner at the most powerful location—the point-of-sale—and to actually exceed the customers' expectations,
many large forest products suppliers and their national associations have
hung their hat on the ISO "certification" process.

Organizations such as ISO were created to serve a different societal purpose than eco-labeling, which involves fundamental independent monitoring
oversight. Industrial-quality, standards-based organizations that try to adapt
their standards to forestry will probably not enjoy the credibility and track
record with environmental groups or consumers that eco-labels have built in
this sector. It is important to remember that while we might know how to
grade lumber or pick apart a sawmill, the forest is the one part of the wood
"factory" where we still have a long way to go before we thoroughly understand its workings. This understanding should not be limited to how wood
grows or is renewed, but must include an understanding of the entire ecosystem, from global climatic effects to managing watershed-dependent fish
populations as a forest product.

For centuries now, forest product companies have made simplistic environmental marketing claims like "we plant two trees for every one harvested." If
an industrial-quality, standards-based organization could successfully pro-
mulgate standards that are acceptable and credible with the public in terms
of creating a high level of resource stewardship, including performance standards that have teeth and are marketable, without destroying their own credibility as highly respected, management standards-based institutions, this approach could succeed.

It might seem that by having two radically different approaches (such as
ISO- and FSC-based certification) consumers will at least be empowered to
compare the approaches and choose the one they feel is most credible. But
with such fundamental differences in approaches, investing in what could
be perceived simply as a more sophisticated self-certification will not only
delay the inevitable but could horribly backfire and set back the industry's
credibility.

As certification becomes more widespread internationally; as consumers,
producers, institutions, and governments become more aware of it; and as the
United States moves increasingly outside North America for more of its forest product needs; the process of certification will become more palatable to
the large companies. More companies trying to gain entry to the U.S. market and The Home Depot will use certification as one of their calling cards
and, as a result, put pressure on North American producers and governments.
I believe this basic tenet will apply to entire countries as well, and even to
publicly owned lands. Competition will drive the certification process in a
most natural way, just as it is doing now.

A company as large as ours has to meet—and want to exceed—the expectations of a public that regards us as a social institution—and severely targets
and punishes us if we do not behave responsibly. Ultimately, the question is
not whether we will have certification, but whether the industry can define
or participate in a credible certification process before certification defines

the industry. Since the issue is not settled as to whether ISO and FSC could develop a common set of principles, it would behoove the industry to acknowledge and participate in the FSC rather than deny its existence, in case an ISO-type approach does not fulfill market expectations. Both ISO's management standards approach and FSC's performance standards approach should be complementary.

Ultimately, the goal of any certification should be to raise the standards for forest stewardship so that the forest products industry is rewarded for taking less, and not more. Instead of fighting product substitution, the industry should embrace it. As long as there is a financial return for an economic activity that society values, these rewards can even result from the worst examples of nonbiodiverse industrial forests being upgraded to old-growth forests, rather than to continually degrade and deforest what remains of our most ecologically valuable forests. Granted, the return of a degraded industrial forest to its former state where it can provide true sustainable benefits can take several centuries.

Certification puts in place the social accounting system that can help these economics emerge. In the long run, I firmly believe that those in the industry with real vision will find certification the biggest financial boon ever. The limits to growth are ominously upon us, with population guaranteed to double to 12 billion souls in the next 100 to 150 years. The diminution of the forest resource needs to be limited. Like the ozone layer, the issue is too dangerous to take chances with. Whether we agree to sustainability by limiting the resource with an OPEC-like strategy, by treaty, or we ultimately have to go to war to save ourselves from ourselves, we need certification as the common accounting system with which we can start the countdown and begin to reverse our self-destructive behavior.

❦ Mark Eisen, Director of Environmental Marketing
   The Home Depot, Atlanta, Georgia

The IKEA Way

IKEA is a home furnishing retailer. It started with a store in a small village in southern Sweden more than 40 years ago. Now there are some 125 IKEA stores in 26 countries. Last year these stores had a total of 116 million visitors. In Sweden we have 13 stores and in the United Kingdom six, so far.

Our largest single market is Germany, with about 30 percent of our total turnover. Germany and the German-speaking countries, the eastern part of Switzerland and Austria, seem to be the most environmentally conscious markets. It was in these countries where—nearly 10 years ago—we could notice the growing customer concern for environmental issues.

The IKEA environmental policy states: "We shall always seek to minimize any possible damaging effects to the environment which may result as a consequence of our activities." This includes raw materials, production, distribution, use, and disposal of used products. It also includes the internal activities in the stores, warehouses, and offices (for example, waste management). One consequence of the policy is that our customers must be able to obtain factual information about the environmental aspects of our products and materials.

To avoid any misunderstanding, I want to emphasize that we are still a long way from fulfilling our environmental policy. We have just started a long process of making environmental considerations part of everybody's work. We have to continuously improve our environmental performance.

Wood is, for obvious reasons, the most important raw material for a furniture company such as IKEA. We use solid wood, particle board, wood veneer, plywood, and MDF board. We also use wood in the form of pulp in cardboard packaging and catalogue paper.

Wood is a renewable raw material with a great potential. Our consumption of solid wood will double over the next few years. Of the solid wood we use, 70 percent is softwood, mainly pine; 65 percent comes from Swedish and Finnish forests.

Until about three years ago, customer questions regarding wood were focused on the possible content of tropical rainforest timber in our furniture. Our answer was that we don't use such timber. The only wood from tropical countries to be found in the furniture range was rubber wood. Recently, we have started on a small scale with teak from Javanese plantations, but only after very thorough on-site investigations and consultation with NGOs.

However, since the Rio conference in 1992 we all (customers, co-workers, management, and suppliers) have begun to realize the importance of good management of temperate and boreal forests. Some NGOs' campaigns have received a lot of attention from the general public, and have activated both buyers and sellers of forestry products. Personally, I am impressed by the actions taken by some of the major Swedish forestry companies during the last few years. I hope that such good examples will be followed by other forestry companies in other countries and on other continents.

An international company like IKEA needs internationally accepted standards. We cannot communicate a large number of different national forestry standards from different supplier countries to environmentally conscious customers, for instance in Europe and North America.

My own background is in quality management. I have looked for some kind of basic world standards for ecologically sustainable forestry—a standard that forestry companies, NGOs, scientists, sellers, and buyers of forest products could agree upon—a standard that also could be communicated in credible and understandable terms to the interested general public.

Last year, together with WWF Sweden and Assi-Domän, we participated in a Swedish pilot project regarding the chain of custody of timber sources. The FSC basic Principles and Criteria, at least presently, appear to be the only available alternative for a world standard for sustainable forest management.

Some North American forest companies apparently want to get a forest management standard included in the ISO 14000 standards series. I do not believe in this approach. The ISO standards (9000 for quality systems and 14000 for environmental systems) are branch neutral. They are not intended for any special kind of material, product, or production process. If the Canadian campaign should succeed, I think it will face large credibility problems with the general public. In that case, what would be the value of such an industry standard?

I see no available alternative to the basic principles and criteria of the FSC. But for the FSC to succeed in practice will require a constructive cooperation and mutual trust between forest owners/companies and NGOs in different parts of the world.

WWF Sweden and the Swedish Society for Nature Conservation have done pioneering work on the Swedish adaptation of the FSC principles. They have consulted with the Swedish forestry companies. It is crucial to the FSC concept that the Swedish work be successfully completed. If we cannot gain acceptance from the major part of forest owners and forest industry here in Sweden, how then will we be able to obtain it elsewhere?

I offer the following challenge to forest companies: Satisfy not only your customers' needs but also the needs of your customers' ultimate customers, "the majority of people." They want to buy a product that does not yet exist, that is, wood furniture made of certified wood from sustainably managed forests. I am not sure the customers are willing to pay much more for this product. But I am sure that in a few years many customers will not pay anything at all for uncertified wood.

I offer the following challenge to environmental organizations: Dare to compromise! You have not failed the environment and the forest issue if you realize that what we need is constructive steps in the right direction now. Do not risk permanent progress on the really important points because of frustrating struggles over details.

Finally, our own greatest challenge is the environmental adaptation of our entire product range. The wood-based products are an essential part of this important task.

❧ Russel Johnson
   IKEA of Sweden
   Almhult, Sweden

B & Q , plc. (United Kingdom)

It is an undeniable fact that as society changes, the culture of our business must change with it—or preferably sooner! The key is predicting that change and being ready beforehand. In the case of the environment, you do not have to go beyond the nearest school to see how much our staff and customers of tomorrow care and know about the environment.

Change is difficult, emotional, and expensive—but it is also exciting and rewarding. This has best been demonstrated by our timber policy. B & Q sells over 250,000 cubic meters of timber a year. This timber comes from over 40 different countries and includes over 40 different species. It is used in products such as joinery, wall paper, and tool handles.

Five years ago, we were asked how much tropical timber we sell and from which countries we buy it. We realized that we could not accurately answer that question. In fact, we couldn't answer it at all. In public relations terms, if you don't know, you don't care. We did care, but we weren't doing anything about it. There was no effective control in the way we bought timber. So in 1990, B & Q recruited a full-time environmental specialist.

There are three important issues in the politics of forest management. First, the forestry problem is not confined to the tropics. Second, a simple boycott of tropical timber is wrong. And third, there is a very clear relationship between the timber trade and deforestation.

It makes no difference if the timber trade is responsible for 1 percent of forest destruction or 99 percent. There is a connection—and that's the point. What concerned B & Q was that there was no way of preventing timber from a badly managed forest coming into our stores. When such timber did come in, business was being damaged, either in the form of customer boycotts, reduced staff morale, lost sales, or bad publicity.

Back in 1990, we had labels on our products. In fact, a quick audit showed that there were over 20 different labels. The problem was that nobody believed them and the sheer number of different types was confusing.

Because of consumer confusion, we removed all of the labels from our stores. However, that didn't improve the fundamental integrity of our sources or give us the proof our customers were looking for. So in September of 1991, the B & Q Board of Directors set a target that by the end of 1995 we would buy only timber from well-managed forests. This was a target set by the World Wide Fund for Nature, and B & Q was the first retailer to sign up to it. The WWF's courage and foresight in working closely with companies has been the foundation of our success, and we believe the 1995 group presented a role model of how other issues could be tackled.

We also set the interim target that by the end of 1993 all our timber would come from known sources. Before that, many of our suppliers, as much as 90 percent of them, did not know or would not tell us which country their timber was coming from. We felt tracing our timber not only to country but to forest level was a vital first step forward. It is really quite basic, for how can we say our timber comes from a well-managed forest if we do not know where the forest is?

To achieve our target, we have developed our own sophisticated process of internal scrutiny. Each supplier's product range is assessed according to three different criteria: (1) the commitment of that supplier, as indicated by their own company policy; (2) the level of detail we have about their precise sources; and (3) our judgment on how well that forest is being managed.

However, internal scrutiny alone is not sufficient to convince our customers that our timber does come from well-managed forests. Internal scrutiny is only self-certification. We need independent certification. Without independent certification, suspicion and skepticism will prevail. Ultimately, self-certification does not work.

We already have certified two product ranges: moldings from our forestry project in Papua New Guinea and Chindwell, and rubber wood exterior doors from Malaysia. Both have recently been re-audited.

There is the potential for a whole host of different certification schemes addressing various forest types, in different ecological and geographical regimes. Some of these schemes may be charitable, some commercial, and some initiated by governments. The end result will be a plethora of different standards, logos, and labels. This will return us to the original problem of too many different labels on our products. We must not repeat mistakes we made in the past, and so must have only one label.

Another problem is how to determine that the certifiers are competent and are working to the same standards as other certifiers. While at first glance this seems complicated and bureaucratic, we do need the certifiers to actually be certified.

This inevitably leads us to the role of the Forest Stewardship Council. The FSC is uniquely placed to identify and promote timber from well-managed forests across the world. No other organization is available today to do that, and that is why B & Q will only recognize certification schemes accredited by the FSC.

Many people still believe that independent certification cannot be achieved, that it is impossible or impractical. We know that is not true. Certification costs of the exterior doors represented no more than 1 percent of the final price and could be much lower.

B & Q's priority now and after the 1995 deadline is to reduce our dependence on internal scrutiny and strengthen our commitment to independent certification. B & Q's greatest role must be to create real commercial incentives for certification. Therefore B & Q will not drop any certified products in favor of noncertified products. We have also set a new target that by the end of 1999 we will stock only independently certified products.

Another contribution from retailers must be to educate the marketplace so that customers recognize and prefer certified timber. From that day onward, a slow but steady conversion of the marketplace will begin and there will be no going back.

The measures I have put in place and the growing awareness of certification leads me to confidently predict that one day B & Q will be trading only in FSC-accredited certified products.

❦ Jim Hodkinson
   Hants, United Kingdom

Sustainable Forestry Certification (Canada)

It is now generally accepted that environmental concerns are here to stay and will continue to be a major influence on business and public policy decisions around the world. In line with this trend, consumers are now beginning to look for some assurance that the forest products they buy are manufactured from timber that has been sustainably managed; certification is a means of providing that assurance.

Sustainable forestry certification then, is simply a voluntary measure whereby producers can verify their forest management systems and credibly present themselves to the marketplace. The operative word here is credibility, a feature provided by virtue of the independent auditing component of the certification process.

Late in 1993, 21 Canadian forest industry associations formed a coalition to address the question of certification nationally and recommend a course of action for industry. Members of this large group quickly came to a common view and embarked on a program that would lead to the development of a certification system for Canada by 1995. It also recommended that international standards for sustainable forestry were necessary and that those would best be developed by the International Organization for Standardization (ISO).

The coalition's view of certification is simple: There is a need for certification, a need that is driven by both public concerns and market forces. Coalition members are in business and will therefore respond positively to the changing situation.

It is also well understood that, in addition to marketplace requirements, there is a growing environmental challenge that must be met, one that is complicated by the fact that worldwide the concept of sustainable forest management is in a period of dynamic change. That process of change is critical to understanding sustainable practices and must be incorporated into any certification system.

Coalition members believe there is a way of doing this—one that uses existing voluntary institutions; one that fully integrates forest certification within the normal business management process; and one that provides assurances to the market that environmentally sound, economically viable, and socially sensitive forestry is being practiced.

The Canadian industry view may be summarized as follows:
Certification is necessary.

- It must be based on the concept of sustainability.
- An international approach is necessary.
- A system of equivalency among countries is basic.
- Compliance must be verified by independent third parties.
- An international framework exists in ISO 14000.
- We should move forward through a multinational, multi-international coalition.

Why Certification?

Over the next few years, the forest sector will be influenced by a range of international initiatives currently in formative stages, including, for example, intergovernmental dialogues and the ISO Environmental Management Program. We expect that many forest industry companies will seek certification under the ISO environmental management systems (EMS) standards, and we maintain that sustainable forest management should be part of that certification.

Of perhaps more immediate concern has been the movement of environmental controversies into the international marketplace. Canadian forest practices in particular have become the focus of attention and are often depicted in a selective and unfair manner. The ability to convince customers that a good job is being done in the forests is no longer just a commercial nicety; it is rapidly becoming a commercial necessity.

Marketplace concerns exist in a variety of forms. There are, for example, threats of boycotts against exporting countries and demands of some retailers for certified wood products. In this perceived vacuum, several organizations are attempting to provide various schemes for verification of sound forest practices. However, without an overall structure to guide standards development, a proliferation of different schemes will only lead to confusion in the marketplace.

For all the above reasons, more anecdotal than empirical, the Canadian forest industry fully concurs with the World Wide Fund for Nature (WWF) that certification is inevitable. Although it may not be for everyone, it is nevertheless a tool that must be available for those who wish to use it.

Sustainable Forestry

Sustainable development is generally understood to be a three-legged stool, with the environment, economy, and society all forming part of the solution. Sustainable forestry and sustainable forestry certification must encompass this model, although it is important to distinguish between intergovernmental agreements and industrial certification.

Intergovernmental dialogues referred to earlier are now producing sets of criteria and indicators for sustainable forestry. These will likely form the basis for an international agreement on forests; clearly, any system of industrial certification should fit within that broad framework. Many of the criteria and indicators, however, are well beyond the scope of any industrial operation; they remain the responsibility of federal or regional authorities and hence are the subject of international, intergovernmental agreements.

Industrial certification, on the other hand, concerns only those factors over which a forest enterprise has some influence. Although different from national criteria, certification can play an important role in determining a country's compliance with its commitment to sustainable forestry.

Standards: The Essential Feature of Certification

The essential element of any certification scheme must be the rule or standard against which a potentially certified enterprise will be measured. To be of value, that standard must be credible, practical in application, and applicable internationally.

If the standards represent such a crucial factor in a certification program, then who is in the best position to undertake their development? There are a number of options—for example, the Forest Stewardship Council, certifying organizations, industry groups, countries, and even customers. However, all these attempts suffer from the lack of one or more of the acceptability criteria referred to above. Again, developing a proliferation of standards will only lead to confusion in the marketplace and ultimately to a loss of public confidence in the principle of certification.

The Canadian coalition has therefore concluded that standards themselves are best developed by the existing accredited and well-recognized, standards-writing organizations—the ISO internationally, the CSA in Canada, and other domestic standards organizations in other countries. These organizations already possess the necessary credibility, expertise, and resources to do the job; it is, after all, their main business. They have a well-earned reputation for integrity and objectivity that arises from a well-defined and rigorous process used in developing standards. This process includes clear guidelines for the representation of all stakeholders who will be affected by the standard, something essential to the long-term credibility of any standard.

In addition, these organizations have in place the international network to facilitate development of compatible standards in a number of countries, a feature necessary to ensure a level playing field in world markets.

Clearly, the development of standards is best left to those who know and understand the process, based on years of experience. That leaves others free to concentrate on the certification process and implementing the sustainable forestry practices, which is, after all, the basic objective.

Sustainable Forestry Standards

Within the framework of environmental management, there are two general approaches to assessing sustainable forestry—product (performance) standards and systems standards.

Product standards focus on the forest estate and evaluate performance on a specific area under management. Like a balance sheet in financial auditing, they represent a snapshot of operations at a point in time. If an audit is successful, it follows that all primary products originating from the area would be certified and labeled accordingly.

Generally associated with this approach is the necessity to maintain a chain of custody throughout the subsequent stages of processing and sale; this is the only way to verify that a final product originates from a sustainably managed forest. However, in an industry known for its complex raw material sourcing situations, it is considered doubtful that a chain of custody can be established and audited at a reasonable cost.

An alternate approach to assessing sustainability is based on the premise that performance is no better than the organization doing the managing. Attention is therefore concentrated on the forest enterprise itself and its management systems rather than the area under management—in essence an application of EMS standards to sustainable forestry. In this case the enterprise is evaluated in terms of its ability to manage in an environmentally sound and sustainable manner.

Management system factors include such elements as policies, clear objectives, planning systems, competent personnel, training, documentation, control measures, public participation, and continuous improvement. Using these criteria will provide evidence of a company's sound, effective internal management, responsibility to the public, and compliance with regulations. They also incorporate the dynamic element of management.

However, systems standards can be augmented with field-level performance indicators for areas actually managed by the organization seeking certification. In a manner similar to financial auditing procedures, a sample of performance measures can be used to verify that reality reflects what the documented system indicates should be in the field.

Coalition members believe that the systems approach should form the basis of sustainable forestry certification, as it is primary evidence of an organization's commitment to sound management over time; it encompasses not only performance indicators, but also the means to ensure that performance is achieved and continues to be achieved over time. This approach is potentially more flexible and inclusive of a broad range of enterprise conditions—which is, in the end, what we are trying to achieve.

❦ Gerald Lapointe, R.P.F.
  Sustainable Forestry Certification Coalition
  Montreal, Quebec

IHPA: A Perspective on Certification

IHPA (International Wood Products Association) represents American importers of wood and wood products from around the world, as well as many others affiliated with the trade. IHPA members strongly support sustainable use of forest resources. Forests are the source of our industry and our families' livelihoods. Our family-owned and operated enterprises have been handed down from generation to generation, in some instances for over a hundred years. Perhaps you can understand the interest in sustainability of someone who wants to pass on a family business to sons and daughters and to see that business remain viable so it can be passed on to their grandchildren as well.

The "Green Certification" issue related to tropical wood and wood products first appeared several years ago, fueled by growing concern about deforestation in tropical, developing countries. Third-party certification of sustainability has been promoted as a means of reducing deforestation by providing wood and wood product suppliers an economic incentive to adopt sustainable forestry management practices.

According to certification advocates, forest management operations that meet the criteria developed by private, third-party, for-profit, certification operations will receive a "green certificate" (for which they must pay a hefty fee) and can promote themselves as being more environmentally friendly than those who don't have a certificate. This will enable them to sell their product at a higher price (the elusive "green premium") and to gain a larger market share due to the unquenched demand on the part of consumers for these "green" products. In turn, certification will allow consumers to feel good about patronizing sustainable products (products bearing the "green certificate") and will assure them that they were doing their part to save the world's forests by forcing drastic changes in forestry management practices.

The appeal of this concept is easy to understand. It offers a seductive and simplistic answer to a very complex problem. However, serious flaws are evident in the logic of certification, and several of its core assumptions must be questioned.

The first problem with the argument for certification is the assumption that global deforestation results from forestry resource mismanagement by the timber industry and that certification can somehow change all that. In fact, it is well understood that in the tropics the primary cause of deforestation is poverty and population pressure. A recent study commissioned by Greenpeace and conducted by the Institut fur Weltwirtschaft, in Kiel, Germany, found that the overwhelming cause of forest depletion is clearcutting of land for agriculture. Forestry was assigned responsibility for only about 2 percent of depletion in Brazil, 9 percent in Indonesia, and zero percent in Cameroon, the three countries specified in the report.

One should remember that harvesting in the tropics is almost exclusively selective, meaning the vast majority of trees are left undisturbed and standing. Alternative land uses, like agriculture, totally remove the trees from the land. Also, only a small amount of the forestry production (typically less than 20 percent) goes into international trade, and the markets where certification is being discussed are small (United States: approximately 6 percent) in the overall international tropical timber trade picture. Using the Greenpeace findings, a quick calculation shows that the U.S. market for tropical timber can be connected to less than eight one-hundredths of 1 percent of tropical depletion (that's less than 0.0008). Thus, it is difficult to see how certification will stop deforestation.

To the contrary, serious restrictions on exports of wood products will jeopardize an industry sector that employs and provides benefits for tens of millions of people in the developing world. Anyone who has witnessed the abject poverty in some of these regions of the world should realize that imposing costly and burdensome trade regulations—even for the well-intended purpose of "green certification"— could jeopardize a very fragile employment base. Is it fair to take such risks with the livelihoods of so many, just so that a few affluent North American and European consumers can assuage their "eco-guilt"? IHPA believes that the time and resources devoted to pushing certification would be better spent on addressing the real problems behind deforestation: poverty and population growth.

Another assumption that must be questioned is that consumers are the driving force behind certification. Are consumers really clamoring for "green" wood products, and are they really willing to pay a "green premium" to get them? Current demand appears to be limited to some who purchase very small quantities and wish to feel "environmentally correct," and to one or two larger buyers looking for a possible marketing advantage. Although a handful of surveys indicate a possible willingness to pay for certified products, the validity of these surveys has been questioned, and other research indicates little consumer interest.

For example, a recent German study conducted by a well-known firm for the Federal Environment Office in Berlin may provide a more accurate estimate of consumer demand for the "green premium." Germany is considered one of the most "progressive" countries when it comes to "environmentalism." The survey involved over 3000 people, the vast majority of whom described themselves as "environmentally aware." While the average level of environmental "awareness" on a scale of 1 to 10 was 7.8, only 36 percent of responders said they would pay any extra for a "green" product, and then no more than a 5 percent premium. How much of the rainforest can be saved with that level of consumer interest?

IHPA has conducted its own informal survey of the market for "green" wood, wood products, and forest management, primarily by monitoring the trade press and other industry publications (including *Import/Export Wood*

Purchasing News, Furniture Today, Woodshop News, and others) for almost four years for signs of a growing market for these products. We have noted only *one* company having placed *one* ad indicating that they were *interested* in purchasing *certified* wood. We also have noted only *one* company whose ad offering wood contained any mention of certification, and that was extremely low-key. IHPA trusts that its members and the readers of these publications know their customers and the business well enough to determine whether there is a market for these goods.

Recently, IHPA was informed that an organization has been formed specifically to create demand for these products in the United States and that this group is receiving significant financial backing from several major private foundations. Obviously, they feel that creating demand where there is no interest is going to be a major undertaking. But, one thing is sure: The argument that certification is a response to consumer demand does not and never did hold any water.

Are some "environmentalists" and certification organizations just attempting to create a market for the certification services they offer? Obviously, and while that may be good marketing, it is not good forestry. What makes these "third party" groups, promoting forest-management criteria heavily weighted toward their social agendas, more qualified to gauge "sustainability" than forestry professionals who have been in the forests for years? Why is the cost for certification so high? Why does it cost at all if the aim of these groups and of this concept is to reward good management? The fees charged by the "independent" certifiers start in the tens of thousands of dollars, even for small family-owned operations, and go up from there. If certification is pursued, *someone* must pay these costs. If even the "progressive" German consumer balks at even a modest premium, someone else must cover the cost: the landowner, the lumber dealer, the distributor, the manufacturer, the retailer? One also needs to wonder about conflicts of interest when the leadership and membership of the "governing body" for certifiers, the Forest Stewardship Council (FSC), appears to be so heavily dominated by those who have either a direct or indirect financial interest in certification. The only clear winners here are the certifiers.

Though certification advocates argue that participation is voluntary, many of them have been attempting to write certification into law. Efforts are currently underway in Los Angeles, in New Jersey, and elsewhere to mandate certification of tropical woods by the FSC for government purchasing. This movement doesn't even follow the FSC line of treating all woods equally: a key point in the rhetoric of certification supporters! Several years ago, the city of Minneapolis turned down a proposal by certification advocates that would have completely banned the sale of uncertified wood in the city. Supporters claimed that it would be the first step toward a statewide ban. Recently, Federal agencies and a number of international organizations have been pressured by certification advocates to support certification and endorsement of the FSC. In response, the U.K. Forestry Commission reflected the thoughts of many who actually work in forest products industry in noting that, with timber-cutting regulations so strict all over the world, the whole certification process seems redundant.

Currently, a number of alternative approaches to sustainable forest management are under consideration, some of which may come to fruition in the very near future. For tropical woods, the International Tropical Timber Organization's comprehensive "Criteria and Guidelines for the Sustainable Management of Natural Tropical Forests" and its "Year 2000" goal have been signed by the governments of the 53 member countries. Similar protocols have been developed for temperate and boreal forest countries (the "Helsinki" and "Montreal" agreements). Also, the United Nations Commission for Sustainable Development created a special "Working Group" to report in 1997 on recommendations for achieving sustainability in the world's forests.

Additionally, national and international standards groups are developing more scientifically based and measurable criteria for forest management. A number of "life cycle" research studies on wood versus competing materials are underway or in development, which may provide a clearer picture of the net environmental impact of wood use. Also, a number of timber-producing countries are very close to implementing "sustainability verification" systems that will be significantly more viable and credible than certification.

Instead of supporting certification schemes, those who are truly interested in promoting constructive and scientifically sound improvements in forestry management should support, with increased funding and technical assistance, professionals engaged in forestry who are today working toward sustainability of forest resources. Research and development work is underway on lower-impact logging, better methods of road construction, and other factors. One project being coordinated by a major international environmental group has even discovered environmental advantages to a form of "clearcutting." Did third-party certification bring this about? NO. Did certification pay for the work? NO. To the best of our knowledge, certification has never planted a tree, and it has never employed large numbers of native people, in the forests, properly managing resources. How much of each dollar gained from certification actually goes back into objective improvements in forestry management and resource conservation?

Can further progress toward sustainability be made? Absolutely. But third-party certification schemes won't bring that about either. While some plantations and some small-scale natural-forest operations have been certified, there is no evidence that third-party schemes even approach practicality in larger-scale natural forest systems.

IHPA strongly supports the sustainable management of all the world's forests. Why? No trees equals no timber. No timber equals no jobs. Sustainable forestry makes good business sense. IHPA, primarily through its C.U.R.E. (Conservation, Utilization, Reforestation, and Education) program, is working both within and outside the industry to raise the level of awareness and increase knowledge of these important issues. From manageable industrial resources to "carbon sinks," to wildlife habitat, and to recreation areas, industry today is acting on its responsibility to make sure future generations can benefit equally from the precious resources that are the world's forests.

❦ Robert Waffle
International Wood Products Association
Alexandria, Virginia

Labor's View

The IFBWW (International Federation of Building and Wood Workers) resolved in its Forest Program, which was adopted at its 19th World Congress in 1993 by 185 unions in 90 countries, that it "supports the labeling of timber from sustainably managed forests." Targeted consumer information campaigns will give timber from sustainably managed forests better access to the market. As a result, incentives will be created for improving the management of forests.

Securing sustainable utilization of forests worldwide, in tropical forests and temperate and cold climate zones, is a big challenge for governments but also for employees and employers in the wood and forestry industries. The realization of this objective calls for securing a framework of conditions for sustainable utilization of forests. This can only be achieved by joint efforts of industrialized and developed countries. The IFBWW therefore demands the following conditions.

• Protection and conservation of forests must be defined as an essential government policy objective; forests must not be treated as a good that is free of charge.

• The necessary legal prerequisites must be created and the necessary financial resources provided for protecting forests.

- The cost of sustainable management must be covered by corresponding income and earnings of the wood and forestry industries.
- Socially just participation by the local population in the earnings from forest utilization must be guaranteed.

Sustainable utilization of forests can be achieved with a low input of capital and technology; but it calls for a high degree of knowledge and skills during planning and implementation. Sustainable and socially acceptable utilization of forests requires the following minimum standards, which must be implemented as binding criteria for certification in the same way as ecological and forest requirements.

Minimum qualifications: Comprehensive training and further training of wood and forestry workers.

Safety standards: Securing adequate occupational safety, health, and accident prevention.

Social standards: Employment in permanent and secure jobs.

Legal standards: The right to form trade unions (freedom of association) and to engage in collective bargaining.

The IFBWW's position is that certification of timber and timber products must include the following principles.

1. Certification of timber and timber products must be done by an independent expert group. The concerned parties, including trade unions, must decide the composition of this body.

2. This independent expert group must have constant access to timber concessions and timber trade houses in order to control the conditions for certification.

3. In the middle term, certification of timber and timber products must be a binding worldwide precondition for market access.

❦ **Comments to the ITTO Working Party on Certification of All Timber and Timber Products**

❦
Certifiers

Scientific Certification Systems

Scientific Certification Systems (SCS) is a multidisciplinary scientific organization based in Oakland, California. Founded in 1984, the SCS mission is to encourage the private and public sectors toward more environmentally sustainable policy planning, product design, management, and production. One major SCS initiative is the Forest Conservation Program (FCP).

An FCP evaluation is structured around three elements that encompass technically sound and socially responsible forest stewardship: timber resource sustainability, forest ecosystem maintenance, and financial and socioeconomic considerations. Clearly, exemplary forest stewardship entails more than sustained timber production. Equally important is the extent to which the integrity of the forest ecosystem is maintained and the extent to which the operation yields benefits to all pertinent stakeholders. To be certified, an operation must meet or exceed threshold standards in each of the three program elements.

Evaluations are conducted according to structured protocols using a three-person scientific team (for example, forester, ecologist, and sociologist or economist) with regional expertise. The forest management audit process clearly defines areas of management strength and deficiency, and establishes baseline performance delineating where companies can make environmental improvements.

All of the participants in the FCP to date have benefited from the information derived from forest management audits and have improved forest management practices as a result. Improved forest management stands alone as the most important achievement of certification. However, certification also allows forest products companies to demonstrate accountability to other stakeholders such as wood products manufacturers, retail buyers, consumers, and government policy makers. Certification provides the mechanism through which forest products companies can communicate, educate, and inform.

Growing Demand for Certified Forest Products

Today's marketplace demands more environmental information about forest products. Independent third-party certification of forest management practices meets that demand. Diverse, market-sensitive forest products companies, like Collins Pine Company (California), Seven Islands Land Company (Maine), Portico SA (Costa Rica), and Duratex (Brazil), have sought certification from the SCS FCP in order to raise credibility with buyers and consumers wary of the timber industry; to differentiate products from the competition; and to gain a competitive advantage based on superior environmental performance.

A series of well-publicized surveys conducted over the past five years have documented the influence environmental problems, such as deforestation and global warming, have had on consumer purchasing habits. Concurrently, distrust of industry's record with regard to the environment has been equally well documented. Market research conducted by The Hartman Group in 1992 showed only 13 percent of respondents believed corporations were "trustworthy sources of information about environmental matters." Independent third-party certification of forest management practices provides consumers with the environmental data they require from a credible, scientific source.

Market Successes of Forest Certification

Anecdotal information from the FCP indicates that certified producers, such as Collins Pine (discussed later), Portico, and Seven Islands, have achieved positive outcomes in opening new markets and obtaining premiums for their wood products. Marketplace successes extend to secondary manufacturers and retailers of certified forest products and cut across all market segments for wood products, from home center retailers and commodity dealers to architecture and design firms, to value-added product manufacturers. Their achievements attest to the challenges and rewards of marketing certified forest products.

Collins Pine directly attributes to certification sales increases of 25 percent to retailers, 22 percent to furniture manufacturers, and 3 to 4 percent to commodity dealers. Portico's certification gained its customers among home centers and retailers such as The Home Depot, Payless, and Cashways. Citing certification as a definite advantage over competitors in rainforest hardwoods, Portico increased its market share in the high-end wood door market by nearly 30 percent in 1994 alone. Certification allowed Seven Islands to leapfrog beyond sales into primary milling and tap secondary manufacturing through chain-of-custody certifications. The company has essentially achieved a vertical integration without making additional investments in downstream facilities. It now receives a 10 percent premium on certified logs and a 5 percent premium on the end value of products such as shingles.

Collins Pine Company, Chester, California, was certified by SCS in 1993 as state-of-the-art well managed. Bill Howe, of Collins Pine, tells why forest certification is important to his company.

> As the forest manager for a mid-sized private lumber company, I believe forest certification can be valuable internally and externally at three levels. At the department level, certification is a vehicle for pursuing excellence. It is a constant reminder of who we are and what we stand for. We routinely make reference to the effect our daily activities have on certification, even though principles already engraved in our management for the past five decades led to that certification.
>
> At the corporate level, certification has provided a third-party resource audit to the landowners. Similar to a financial audit, certification assures landowners who are absent from the daily landscape that their resources are being managed in a way that maintains environmental integrity and resource sustainability.
>
> At the regional level, it is our hope that acceptance of certification grows within the forest products industry so that it raises the industry's perception of environmental values equal to the forest's financial values. Whereas we have long argued against stifling governmental regulations, certification may provide the internal desire for industry to modify its own goals and voluntarily achieve results that regulations may rarely accomplish.

Conclusions

Third-party certification offers many potential benefits to forest products companies: improving forest management practices that increase the value of the land—the largest single asset of any forest products company; raising employee morale and enhancing the ability to recruit the best-qualified staff by affirming a commitment to long-term forest stewardship; improving government relations; and increasing a company's access to public officials. Ultimately, the end consumer will determine the degree of certification's marketing success. However, for many certification participants and especially for companies producing forest products, the benefits derived from improvements in forest management practices alone justify the industry's commitment to certification.

❦ Debbie Hammel
   Director, Forest Conservation Program
   Oakland, CA USA

❦
Geographic Regions

Latin America

Who Should Be the Certifiers?

Most certifiers today are from industrial countries, and almost all dedicate themselves wholly or partly to activities in tropical countries, according to rules written by themselves. Some pretend to become "multinationals" of the certification business. In most cases, and despite the overwhelming attention to tropical countries, there is little to no participation of experts from these regions in such organizations.

Nevertheless, there is a growing awareness among tropical countries that they must control the certification process in their own countries themselves, that appropriate certification organizations should be set up in their countries to undertake this responsibility in an unbiased manner, and that for certification organizations to be legitimate, they must abide by certain rules, and be accredited by an objective and representative body. Only then will they earn the trust of consumers and achieve market credibility.

Certifiers should be objective, independent, and with suitable infrastructure and human resources to carry out their responsibilities with an acceptable degree of precision and credibility. They should preferably be structured according to guidelines already established by the International Organization for Standardization (ISO) for certifying organizations, such as the European Norm EN 45011. Should this be the case, accreditation would take place within a national regulatory framework, would be established under a common set of rules, and would facilitate trade through the international recognition of different certification programs.

Implications of Certification

At present a very large proportion of the production of industrial timber in Latin American countries comes from forests without any form of forest management, from 50 percent in Venezuela to 80 percent in Peru. Most of this production is associated with the expansion of the agricultural frontier or with illegal logging.

The rest of the production, from areas under concessions and subject to management plans, may for the most part have serious difficulties in meeting the requirements of the ITTO guidelines or the FSC principles and criteria.

Latin America produces as much industrial timber as Southeast Asia, about 110 million cubic meters a year. But contrary to Southeast Asia, where exports account for about 40 percent of production, largely in the form of logs, in Latin America only about 10 percent is exported under different levels of processing. Logs represent a negligible proportion of total exports.

It may seem, then, that as long as certification is restricted to international markets, the regional effort necessary to meet such requirements is insignificant. In fact, the proportion of production reaching the international market is an integral part of total output; it is difficult to make a practical separation between the two markets. Furthermore, the need to base the local market on sustainably managed operations is far more important for the maintenance of the resource than the international market.

On the other hand, the commitment (ITTO year 2000 objective) made by tropical countries to sustainably manage their forest by the year 2000 is also the commitment of industrial countries. There is an inherent obligation on their part to assist tropical countries to reach that objective.

It is widely recognized that sustainable forest management is associated with the degree of transformation of the raw material (logs) of the country of origin. The need to produce and export finished or semi-finished products, create jobs, and retain added-value is embedded in the ITTO objectives. Further processing is associated with industrial development, which in turn is tied to access to technological and financial resources. Industrial development can take place only if there is unobstructed access to the main international markets, and depends on the availability of an adequate infrastructure of roads, ports, energy sources, and a trained labor force. Sustainable forest management is thus associated with a chain of requirements that needs to be considered in national forestry development plans, as well as in the position of tropical countries in international negotiations.

One of the most quoted concerns about certification is its cost, and the degree to which that cost is recognized as part of the price of the products in the markets. All practical indications are, however, that for middle to large producers in the region, certification represents a minor proportion of the commercial value of the raw timber produced. At the higher end of the spectrum, the cost of a certification operation may be in the order of U.S.$50,000. This represents from 1 to 10 percent of the yearly production costs of average industrial operations in the region, ranging from 10,000 to 50,000 cubic meters of roundwood per year. When compared to the market price of the product, the cost of certification would tend to range from 0.5 to 4 percent of the total commercial value of the log production. All calculations seems to indicate that price changes due to certification ranging from 10 to 15 percent could be transferred to the market, in return for the supply of timber proven to originate in forests managed according to internationally recognized standards.

For smaller producers, cost may represent a larger and possibly prohibiting proportion of the cost of production. For these cases, a system of subsidies or aid would need to be established, including a reduction in costs and the simplification of the certification process itself.

Certification and Eco-labeling

Certification would apply to forest management and not to the products in the market. The labeling of products becomes an additional operation, which may have different implications. A label could simply attest that a particular product comes from a forest that has been certified as sustainably managed. Or it could be an eco-label, with much wider implications, where the processes of manufacturing, transportation, use, and disposal are also taken into account as ecologically sound.

In the long run, the tendency seems to be for all wood products to be eco-labeled as environmentally sound, taking into account their full life cycle, from the forest to their disposal after use. In the short run, however, the tendency is to focus on the certification of forest operations, according to universally agreed-upon standards, and the labeling of products as coming from those forests.

(From an invited paper to the workshop on policy reforms for the conservation and development of forests in Latin America, sponsored by CIFOR, the World Bank, USAID, and IICA. Washington D.C. July 1-3 1994.)

❦ Julio César Centeno
Mérida, Venezuela

A Southern Perspective

Certification of timber from well-managed forests is a new but promising mechanism for promoting forest conservation. For consumers, it provides a means of rewarding forest owners for sound management practices and ensuring that the woods they use do not harm the environment. For producers, certification offers greater access to markets and higher prices. For conservationists, it provides a market-based tool to affect forest management. For governments, certification can help to ensure that forests are not over-exploited and that the resource needs of their future citizens will be respected.

Certification still has many challenges. One important issue that has received very little attention is the imbalance between the South (developing countries) and the North (developed countries) in the structuring of certification schemes.

At present, almost all certifiers are based in Northern industrialized countries, while many, if not most of the operations certified are in tropical countries. This has important repercussions for the distribution of wealth, the development of human resources, and the forging of international consensus.

Certification is a costly enterprise. Rainforest Alliance, for example, charges $10,000 for a small- to medium-sized inspection, in addition to overhead and transportation costs. In another case, SGS Forestry in England charges about $30,000 annually to inspect a 500,000-hectare timber concession in Guyana. I do not wish to debate these costs here but simply to point out that they can be significant, especially as certification becomes more widespread.

There are several good reasons to be concerned about these high costs and inequalities. In the first place, certification is intrinsically linked to sustainable development. Certification schemes emerged from the realization that forest management has an important role in a global conservation strategy and that consumers should not expect forest owners to make altruistic decisions.

Without a balanced participation of southern certification organizations, tropical countries will increasingly react against the process. Northern domination fuels the arguments of those who see certification as a threat to developing countries, and it undermines those who consider certification to be a catalyst for change and part of a broad strategy for forest conservation and sustainable development.

The high cost of certification may limit its development and reduce the share of any "green premium" to reach forest owners. If certifiers have to fly in and out of tropical countries and are paid international consultant's fees, the cost of certification will be much higher than if it were conducted on a regional or national level. Consumers will fund the overhead costs of northern certification organizations and consultants rather than supporting forest owners and managers in developing countries.

As part of a global strategy for sustainable development, certification can be expected to follow new paradigms and higher ethics than have traditionally governed North/South relations. Poverty does not foster a long-range perspective. In poor households and communities, land-use decisions understandably respond to immediate needs more than to the needs of future generations. As people and societies secure their basic needs, they become better able to consider long-term sustainability in their resource allocation decisions. Therefore, as proposed by the United Nations Commission on Sustainable Development, reducing inequalities is a goal of sustainable development. In fact, the dominance of northern certifiers poses a threat to the efficacy of certification as a mechanism for forest conservation.

Development of human resources is a key to promoting sustainable forest management. Certifiers are in a unique position to acquire expertise in sustainable forestry. This expertise is unlikely to be shared by international consultants on a tight schedule and is more likely to remain within the institutions and countries where the certifiers are based. This is a serious problem because certification is only one aspect of sustainable forestry, and a relatively simple one. Managing forests on a daily basis is a much greater challenge. Without well-trained foresters and technical support staff in the developing world, sustainable forestry will never become a widespread global land-use alternative. Obviously, certifiers cannot undertake the responsibility to train foresters from developing countries; but, as much as possible, they should take training into account. By supporting certifiers in developing countries and encouraging trainees in sustainable forestry, certification can make an important contribution to developing these essential human resources.

By acknowledging the dominance of northern certifiers at this early stage in the certification process, we have the opportunity and the responsibility to change it. One strategy is to encourage the development of certifiers based in tropical countries. This could come in the form of financial support and technical training. Given the rapid dynamics of certification, we need an affirmative-action policy to train southern certifiers in the latest developments and standards. To increase access to information, certifiers should actively participate in technical and scientific meetings and extensive networking. Networks could amount to a simple newsletter or an electronic bulletin board.

We need broad international cooperation to develop consistent certification standards with scientific rigor. Most forest ecosystems still lack detailed management standards based on solid science. Truly cooperative programs—not "parachute science"—can be an important complement to certification initiatives. Partnerships between certifiers from different countries would promote the exchange of information and expertise and would help build institutions and equitable relationships. Such partnerships would also give northern certifiers increased international recognition and credibility.

Greater participation by local and regional certifiers can reduce the risk of inappropriate certification. In tropical countries, social problems are major obstacles to sustainable forestry. A certifier is much more likely to overlook a serious social problem if he or she only drops by occasionally from outside the region. International certifiers operating in regions where FSC national or regional initiatives and working groups exist should be accredited by those bodies.

The call for greater participation of certifiers from the developing world should not be romanticized. International inspectors will be needed to ensure that different regions adopt similar standards and that these standards are enforced. This is exactly the role of the Forest Stewardship Council: to accredit and monitor certifiers. In all of these activities, our goal should be to strike a fair balance between northern and southern participation.

(Adapted from WARP *Understory* 4(3), Summer 1994.)

❦ Virgílio M. Viana
   University of São Paulo, Brazil

Southeast Asian Perspective

Exploitation and export of Southeast Asian forest products began in earnest after World War II in response to demands from Europe. The region is presently the world's leading producer and exporter of tropical timber products, with the bulk originating from Indonesia and Malaysia. It is expected to remain a major exporter with sustained production from Indonesia and Malaysia, increasing production from Cambodia, Laos, and Myanmar, and growing investments in forest plantations.

Most countries in Southeast Asia have strong foundations for sustainable forestry. They include a well-established forest service, sound forestry legislation, pragmatic forestry policy, and, above all, an extensive forest estate dedicated to sustainable management of the environment, biological resources, and harvest of all types of forest products. Most of the countries are committed to the ideals of international forestry organizations such as the FAO and ITTO.

Certification / Eco-labeling Issues

Most countries in Southeast Asia have agreed to ITTO's guidelines and criteria for sustainable forest management as well as its Year 2000 target. The growing demand for independent certification/eco-labeling at an earlier date, which is seen as inconsistent with ITTO's objectives, has created considerable confusion and a dilemma for timber exporters. The situation has been aggravated by the proliferation of certification/eco-labeling schemes and unilateral actions against tropical timber.

The need for certification/eco-labeling to verify sustainable forest management has been accepted in principle in Southeast Asia, but its effectiveness for achieving that objective has been questioned. Nevertheless, Indonesia has developed its own eco-labeling scheme, based on the ITTO, WWF, and FSC requirements, and established the Indonesian Eco-labeling Foundation/Institute to implement it in 1995. Malaysia is preparing for timber certification/eco-labeling and plans to establish a National Timber Certification Center, but maintains that labeling should:

- be applied to all types of timber and substitutes for timber.
- be based on internationally agreed-upon standards and criteria of sustainable forest management for all types of forests.
- be introduced within a realistic time frame for achieving sustainable forest management.
- most importantly, not be allowed to operate as a nontariff barrier but rather to serve as an incentive for unrestricted free and fair trade of all types of timber.

Conclusions

Most countries in Southeast Asia have the basic foundations for sustainable forest management and should be able to comply with the requirements for certification/eco-labeling. However, many believe that additional human and financial resources would be necessary for compliance, which could render tropical timber products noncompetitive. There are also considerable lingering doubts about the real motives of certification/eco-labeling and its effectiveness in achieving sustainable forest management.

The prevailing confusion and uncertainties surrounding certification/eco-labeling of forest products in Southeast Asia is likely to impede the achievement of sustainable forest management in the region if the misperceptions and issues are not clarified. It is imperative that concerted actions be initiated to correct the impression that certification/eco-labeling is driven by vested interests and is discriminatory. The active involvement of all stakeholders, including the forest products industries and exporters, is inevitable if certification/eco-labeling of forest products is to gain wide acceptance in Southeast Asia. Strategic alliances or smart partnerships between relevant stakeholders should be formed to promote certification/eco-labeling as a legitimate and effective means for achieving sustainable forest management.

❦ Sian Tuan Mok
Kuala Lumpur, Malaysia

❦
Multilateral Agencies and Funders

The World Bank and Forestry

In the following discussion, when reference is made to "the Bank" the term refers to the World Bank Group that is made up of two lending arms, the International Bank for Reconstruction and Development (IBRD) and the International Development Association (IDA), that lend to governments of member countries; a risk guarantee agency (MIGA); and the International Finance Corporation (IFC) that lends to the private sector in member countries.

After extensive review of its 40-year experience in providing development assistance to the forest sector in a wide range of countries, and consultation with stakeholder groups, the World Bank adopted a new Forest Policy in 1991 (*The Forest Sector: A World Bank Policy Paper.* Washington, D.C. 1991). This policy seeks to promote sustainable and conservation-oriented forest management and lays out five principles that guide the Bank's involvement in the forestry sector. These are:

1. Assisting governments to better manage the intersectoral linkages that affect forests.

2. Strengthening international cooperation to promote sustainable forest management and conservation.

3. Promoting policy reform and institutional strengthening to rectify market and policy failures that encourage deforestation and inhibit sustainable land use. In this regard, special emphasis is placed on expanding participation in forest resource planning and management and mobilizing private sector resources and skills.

4. Expanding and intensifying the management of areas suitable for sustainable forest production, including the establishment of plantations to reduce pressure on the existing forest resource base where the scope to do so is sound from social, environmental, and economic perspectives.

5. Preserving intact forest areas by supporting initiatives to expand the designation of forest areas as parks and by adopting and encouraging borrower governments to adopt a precautionary approach to forest utilization, particularly in tropical moist forests. In this regard, the Policy makes the specific commitment that the World Bank Group will not under any circumstances finance commercial logging in primary moist tropical forests.

Independent certification of forest management and wood products is potentially a powerful tool to encourage widespread practice of sustainable and conservation-oriented forest management that the Bank's Forest Policy seeks to promote. This section examines the opportunities for the Bank to encourage timber certification initiatives within the framework of its 1991 Forest Policy, but first highlights critical constraints to it playing a more central role in promoting them.

Constraints to World Bank Support

There are three main constraints that limit the World Bank Group's ability to promote or support forest management certification initiatives. First, while the Forest Policy of 1991 promotes many of the same principles and objectives as certification, it also imposes a significant limitation on the Bank's involvement because the Policy commits the Bank Group to not financing commercial logging of primary moist tropical forest—a tenet that it has upheld. There is a broad range of opinion, within and outside the Bank, as to how this clause should be interpreted. Some believe that the Bank should not finance preparation of management plans, upgrading of management systems (including introduction of reduced-impact logging systems), or any other activity that contributes, in the end, to cutting trees. Others believe that those investments should be made by the Bank because, ultimately, they improve the quality of forest management, reduce environmental damage, and promote economic development—and thus contribute to the Bank's ultimate goals.

Second, the Bank is primarily a lending institution. Most of its loans are extended through the IBRD and IDA directly to governments, to finance development projects. These agencies cannot lend directly to regional governments, national NGOs, community groups, or private companies. For the forest manager, the costs associated with obtaining forest management certification are those of (a) having the operation assessed by a certifying company, according to defined certification criteria, and (b) upgrading forest management practices to achieve certifiable status. If the forest owner or manager is a private company or community group, neither the IBRD or IDA can provide financing for these costs. If, however, it is the State, the Bank can provide funds for introduction of better forest management practices, as long as the Bank's Forest Policy is not violated. Many governments are reluctant to borrow for such purposes, however. And while the IFC can provide finance to private companies that wish to borrow for this purpose, it is also governed by the Bank's Forest Policy.

Notwithstanding this limitation, there are other ways in which the Bank can provide or facilitate use of grant funds to support certification. Although the Bank has very few special grant funds, none of which are specifically dedicated to the forestry or natural resources sectors, a grant of nearly $450,000 from the Institutional Development Fund (IDF) was made to Indonesia to help establish the Lembaga Ekolabel Indonesia (Indonesian Institute for Eco-labeling). Further use of this fund by other countries is possible. Alternatively, many Bank projects are co-financed by grants from bilateral agencies or are closely associated with projects financed by Global Environment Facility grants. Instead of using a Bank loan or credit, these grant funds could be directed toward certification-related activities that the Bank cannot fund directly, or to support NGOs or the private sector directly.

A final but significant limiting factor to the Bank's support of forest management certification is that, in reality, relatively few timber-producing countries in the developing world are likely to introduce certification on a national scale in the immediate future. Those that are most likely to do so are countries that depend heavily on North American and European markets for their exports, where consumers' environmental concerns are bringing about changes in timber-buying habits, away from noncertified timber. In most tropical timber-producing countries the majority of timber (the global average is 75 percent) produced is consumed domestically; there is not as yet sufficient environmental awareness in those consumers to demand sustainably produced timber. The immediate opportunities for Bank involvement may, therefore, be quite limited.

World Bank Contributions to Certification Initiatives

There are three levels at which mechanisms are needed to effectively introduce a global system of forest management certification—global, national, and field levels. At the global level, the Bank can contribute significantly to the debate on sustainability and timber certification by undertaking research and drawing on its worldwide experience to provide incisive analysis on its feasibility and likely impacts, and disseminating the results widely. Two papers (see references) have been published so far on the economic and financial implications of timber certification and its likely impact on trade as well as several others on various aspects of forest management and conservation. At the field level, the Bank (through IBRD and IDA) is quite limited in its ability to fund certification-related activities—such as introducing new harvesting equipment, drawing up detailed management plans, or training personnel of private forest management firms—due to the Policy and lending constraints described above.

The level at which the Bank can provide the strongest support is at the national level. It is at this level that ongoing activities of Bank-financed projects are most congruent with forest management certification initiatives and objectives. The Bank could help to more readily facilitate the creation of national certification systems, standards, and procedures by placing greater emphasis on particular aspects of policy analysis, policy reform, and on well-targeted, project-financed activities.

Support through Policy Analysis and Reform

During recent years the Bank has increased funding of forest sector reviews (FSRs). FSRs have recently been completed in Mexico, Argentina, Costa Rica, Bolivia, Malaysia, Indonesia, Nigeria, India, and Zimbabwe; another is underway in Colombia. A key objective of these reviews is to identify ways in which governments can strengthen their commitments to sustainable forest management by making specific changes to sectoral policies—by eliminating policy and market failures, for example.

A particular benefit of these analyses is that they provide a comprehensive overview of all aspects of the sector. This enables the country, the Bank, and other donors to identify areas in which support is most needed. They could also be used to prioritize actions to accelerate introduction of national certification systems and development of improved forest management standards. For example, outdated policies and laws that hinder development of, and compliance with, sustainable forest management principles could be redrafted with Bank support. Introducing better policies would thereby make it easier for forest managers to comply with field-level criteria of forest management certification. Further examples include the introduction or revision of laws to provide for:

- Creation of an adequate representative network of protected areas at the national level
- Greater local participation and benefit capture in forest management and policy making
- Revision of taxation, royalty, and other revenue-generating mechanisms to begin to internalize the costs of (environmental) externalities
- Incentives to encourage companies to upgrade their field management practices
- Higher environmental standards through implementing national regulations for forest operations, such as forest retention targets for biodiversity conservation within production forests, or the maximum slope allowed to be logged
- Social, economic, and environmental assessments of forest management operations
- Titling or regularization of forest land claims ensuring full rights to traditional forest dwellers and users

Support through Project-Level Activities

Every country that pursues forest management certification will need to establish an institutional structure and mechanisms to undertake several functions: to set nationally appropriate criteria and indicators for sustainable forest management; to accredit and regulate certifiers; to monitor compliance with certified status; to impose strict fines or other penalties on companies that are awarded accredited or certified status but violate the conditions; and to coordinate with other countries, international certification bodies, timber wholesalers, retailers, and consumers. While in many countries this management system may be an entirely nongovernmental structure, in others governments may choose to be involved to some degree—depending on the extent of their ownership of forest land and direct involvement in forest management. Some governments may wish to regulate the whole process; others may prefer the process be managed by an independently appointed and regulated body comprising representatives of all interests in the sector.

Because its primary interlocutor is government, the Bank is better placed to provide financial and technical assistance to those governments that wish to encourage, or be closely involved with, regulating or implementing forest management certification. Areas in which the Bank could increase its support at the national or subnational level to contribute to improved forest management (whether or not certification systems are introduced) include:

- Development of nationally appropriate forest stewardship standards for each ecosystem or forest type
- National or regional inventories to lay the groundwork for detailed forest management planning
- Training and extension programs for government and private company staff, and local people, in better forest management techniques
- Development of applied research into appropriate forest management techniques in different forest types to inform, refine, and establish forest management standards
- Design and introduction of efficient timber-tracking systems to facilitate greater forest revenue capture and enable identification of sustainably produced wood in the market
- Creation of private, independent forest management inspection and monitoring services

Thus, while the Bank Group is not directly involved in many forest certification initiatives at present, there is certainly scope for it to play a stronger part in the future.

References

Varangis, N. V., R. Crossley, and C. A. Primo Braga. *Is There a Commercial Case for Tropical Timber Certification?* Policy Research Working Paper No. 1479, June 1995.

Varangis, P. N., C. A. Primo Braga, and K. Takeuchi. *Tropical Timber Trade Policies: What Impact Will Eco-labeling Have?* Policy Research Working Paper No. 1156, July 1993.

❦ Rachel Crossley
World Bank, Washington, D.C.

The World Environment and Resources Program

All over the globe, natural forests are disappearing at a rate that is unprecedented in human history. The demand for timber and nontimber forest products is growing steadily and seems unquenchable. Continuing expansion of the world's economy, projected population growth, and the essential role of forest products in human subsistence all contribute to that demand curve. The consequence of this widespread destruction of boreal, temperate, and tropical forest ecosystems goes far beyond currently perceived shortages of world timber supplies, increased housing costs, or lost jobs in the forest industry sector. Innumerable species and habitats are disappearing, and our natural resource options for a sustainable future existence are diminishing.

Conservation and Commerce

The World Environment and Resources Program has supported dozens of local community-based programs attempting to demonstrate sustainable forest management throughout the tropics. While sustainable forest management is still largely a marginal concept compared to conventional ideas about forestry, traditional industry practices, and long-established public policy, it holds real promise of advancing both conservation and economic development objectives. Sustainable forestry aims to provide a continuous yield of high-quality forest products while preserving the essential biological and ecological integrity of a healthy, self-perpetuating forest.

These new approaches for sustainable forest management are, however, still young and rarely used. They remain on the margin of mainstream research and economic development priorities. Part of the problem shared by our grantees at the local level is that the leaders of most projects involved in these experiments suffer from inadequate understanding of the broader context or overall system of market linkages and resource flows in which they must function.

In response to these barriers to the widespread implementation and commercialization of the concept of sustainable forest management, a small working group, supported by the MacArthur Foundation, met during the past year to describe the forest product market system in a simple way and to identify opportunities or pathways that could accelerate needed transitions in the marketplace. The five team members brought a variety of experience in the forest products industry, resource management, the certification movement, "green" marketing initiatives, the policy and regulatory environment, and capital markets. Challenged to devise a more systematic approach to the Foundation's investments in sustainable forest management, the working group reviewed critical aspects of the concept, including production and harvesting, monitoring and certification, trading and manufacturing, and education and promotion. Their primary goal was to gain a clearer understanding of how to improve the likelihood of achieving sustainable forest management.

The working group's major output is a framework for understanding the current market system in the form of a multilevel, multifactor matrix. While this work has received considerable attention within the forestry community, both domestically and internationally, its primary purpose was to inform Foundation staff. We intend to use this matrix to help us improve our ability to identify strategic interventions, not only for the World Environment and Resources program, but also for the Foundation's loan window and placement of capital investments.

The grants approved by the Board in March of 1995 under the heading of Conservation and Commerce represent the beginning of a more systematic approach to this critical set of issues. Certification is one of the set of market mechanisms that we are supporting which offers unique opportunities to have a major impact on market trends in the forest products sector worldwide.

The award for the certification program of Rainforest Alliance's Smart Wood Project activity in Latin America gives us an opportunity to encourage market forces, through wood products certification, and to shape consumer and producer behavior. Designed to evaluate the performance of particular forestry operations, wood certification has been gaining experience and gathering steam since its inception in the late 1980s. In theory, certification could reward producers with reliable supplies, new markets, premium prices, and higher profits. With three years of successful U.S. regional assessments and certification, the Smart Wood program with MacArthur support now plans to expand its activities in a number of critical southern countries including Brazil, Mexico, and Costa Rica.

❦ Michael Jenkins
John D. and Catherine T. MacArthur Foundation
Chicago, Illinois

Key Issues on Certification of All Timber and Timber Products

Following is a summary of key issues that emerged from discussions of the ITTO Working Party in Cartagena de Indias, Colombia.

1. Participants in the Working Party expressed a full range of views on timber certification. Many governments have not adopted a formal position on timber certification and are continuing to study the issue. Further discussions on this issue should be considered in their appropriate fora.

2. Some participants felt that the variety and inconsistency of labeling and certification proposals and initiatives are causing concern and confusion.

3. Any certification scheme should be compatible with relevant international obligations related to both trade and environment and should respect national sovereignty.

4. Certification schemes should be consistent with the UNCED global consensus on sustainable forest management.

5. Certification schemes should be internationally recognized and based on comparable guidelines and, to the extent practicable, on scientifically based criteria for sustainability of timber from all types of forests. This may call for harmonization to ensure compatibility and serve as a basis for mutual recognition.

6. Some participants considered that certification is one of the market-oriented tools that may serve as an incentive to promote sustainable forest management and that should function as a complementary element of a broader policy package.

7. It is important to assess the potential positive and negative impacts of the certification of timber, and the potential for causing the substitution of nontimber products for timber products. It is still unclear how the market will respond to whatever additional costs and benefits may be associated with certified timber.

8. Certification schemes should be voluntary and based on transparent systems developed on the basis of broad public participation.

9. Timber certification schemes should be neither discriminatory nor the basis for trade measures incompatible with the GATT. Several participants emphasized that the scope of certification should cover all types of timber. It was also stated that comparable arrangements for timber substitutes should be considered in other appropriate fora.

10. Several participants expressed the view that a workable timber certification and labeling scheme will require a realistic and suitable time frame for implementation, which shall not be sooner than the agreed objective by which the sustainability of all types of forests is to be attained. Other participants noted that an increasing number of voluntary certification schemes are currently being developed and implemented; and where sustainable management practices have been implemented, voluntary certification schemes deserve to be encouraged.

11. Participants from ITTO producing member countries stressed that for any certification schemes to be credible there is need for cooperation with developed countries for the attainment of sustainable forest management through the transfer of technology on preferential and concessional terms, new and additional financial resources, human resources development, and institution building.

12. Some participants from ITTO consuming member countries suggested that certification schemes can be financed through market mechanisms.

13. Some participants suggested that certification schemes should be available for all scales of operation.

Responses to ITTO Report

The ITTO Working Party on Certification of All Timber and Timber Products convened in 1994 in Cartegena de Indias, Colombia. Following are excerpted responses of selected governmental delegations to the ITTO report, taken from Appendix II: "Text Statements by Participants."

Brazil

The Brazilian government is of the view that timber certification or labeling should not be required before developed countries have fulfilled their commitments to provide new and additional financial resources and to transfer the necessary technology to enable developing countries to attain sustainable management.

We believe that certification could only be adopted on a voluntary basis and for all types of timber. As a result, the Brazilian Government is of the opinion that the ITTO, for dealing only with tropical timber, has no role to play regarding timber certification or labeling. In our view, GATT—or the WTO that will succeed it—is the appropriate body to discuss the issue.

United States

The U.S. delegation notes that certification and labeling systems of a private and voluntary nature promote efficiency and avoid difficult questions that would be raised by government-sponsored or mandatory systems. The United States supports the use of private and voluntary mechanisms within the framework of existing legal systems.

Nevertheless, there continue to be reservations about the use of such systems to promote sustainable management of forests. In addition, many questions remain about the proper design and operation of such schemes, including questions related to their utility, their costs and benefits, their effectiveness, the validity of their criteria, their operations, and their relationship to trade policy.

Indonesia

Indonesia recognizes that, to be workable, the timber certification must be based on internationally accepted criteria, indicators, and standards.

While we recognize the importance of the external intervention through forestry auditing in the future timber certification scheme, we are of the view that it should be kept to an appropriate minimum level. We would prefer to have a certification scheme that is home-based, using local personnel for inspection, assessment, and so on.

Japan

We view timber certification as being related to a general question of how to harmonize trade and environment policies. Certification requires careful consideration in the context of free trade principles as well as compatibility with GATT.

We feel that many questions should be answered. Particularly, there must be convincing evidence that certification and labeling will indeed promote sustainable forest management.

We are aware that timber certification and labeling are insufficient to deal with the whole issue of sustainable management. With regard to the argument that timber certification and labeling should be applied to all types of timber, we agree with the points of the argument, but in principle, ITTO is a forum with the mandate for tropical timber only.

Malaysia

In principle, Malaysia can appreciate the rationale that timber certification and labeling is being advanced in the name of sustainability. In the event that certification and labeling of timbers is deemed desirable and feasible, it is essential that it be made applicable to all types of timber, while similar and comparable arrangements should be considered for timber substitutes, including plastics, aluminum, and steel.

Malaysia is concerned over the proliferation of unilateral proposals on timber certification and labeling. Ideally, all existing unilateral efforts should be discontinued and replaced by a single global mechanism that applies to all timbers.

Sweden

What is sustainable management varies between different types of forests and areas. Thus, in our view, sustainable management should be defined on a local or national basis. Global criteria are indeed desirable but could not be directly applicable at the local level without modification.

At the international level, in consequence with this preference for local or national definitions of sustainable management, my government, industry, and other interested parties in Sweden would prefer an independent, voluntary, and market-oriented organization to coordinate the different sustainable management schemes.

United Kingdom

The U.K. is working within the E.U. on eco-labeling processes often involving very complicated "cradle to grave" analyses. We have sympathy for the view that certification should look at products—for example, window frames—made from all materials rather than just timber.

We believe schemes need to be voluntary and negotiated between producers, whether government or private sector, and the traders, importers, and retailers. For credibility, it seems inevitable for NGOs to be involved in the process.

Congo

We see certification as originating with the needs of consumers in Europe rather than being a tool to promote better forest management. However, in the end, consumers and producers of timber have a common concern—the sustainable use of forest resources.

Sustainability is a complex and poorly understood concept, at least in tropical forests. There is a need to improve forest management on the ground first before obtaining certification—how are developing countries going to secure the resources needed to do this? If certification is to be accepted it must apply to all types of timber.

Peru

We are currently finalizing new forestry legislation that requires implementation of management plans with comprehensive criteria for sustainability. We agree with the objectives of certification as long as it applies to all types of timber. We are not sure how certification can be applied in practice, however, and believe that this needs further study. The costs and benefits of certification should be studied carefully to ensure that small-scale producers, for example, are not discriminated against.

❦

Conclusion:
Prospects for Certification

The previous parts of this book indicate that certification is a complex issue that can be viewed from many angles, and that different stakeholders have distinct perspectives. Here we attempt to draw some general conclusions and explore alternative scenarios for the future of forest certification.

Key Issues

First, there seems to be a growing consensus in many regions that certification is an inevitable phenomenon in forestry. It is interesting to note how rapidly the attitude of many stakeholders has changed from questioning *whether* certification is going to happen, to *how* it will happen.

Second, the attitude toward certification has gradually changed from distrust and suspicion to growing interest and active engagement from stakeholders ranging from social movements and environmentalists to business leaders. This can be partly attributed to the educational process that has resulted from the growing number of symposia, workshops, and other meetings on certification. It can also be attributed to the potential of certification for addressing some of the challenges of forestry today.

Third, the impact of certification varies and is likely to continue to vary among regions. Exporters producing for markets in Europe, North America, and urban centers of Latin America are likely to become increasingly involved with certification because of consumer awareness and interest. In contrast, producers geared toward other markets such as East Asia, are likely to be less involved in certification processes.

Fourth, certification needs to maintain a worldwide focus to secure international acceptance. Although it originated as a tool for promoting tropical forest management, certification needs to maintain a broad scope because the challenges of forest management are present in all forest ecosystems of the world. In addition to being unacceptable to many constituencies, a geographically biased focus would have little scientific justification.

Fifth, certification represents a refreshing input of new concepts and perspectives for the forestry profession. Foresters have pioneered the concept of sustainability by developing protocols to evaluate and promote sustained timber yield from forest stands since the 18th century. Now the challenge is to take another step and add a greater emphasis on socioeconomic and environmental dimensions of forest sustainability. This requires not only a greater interdisciplinary focus, but also the ability to work with consensus building and participatory multistakeholder planning of resource use. Overall, certification represents a historical opportunity for growth of the forestry profession as a whole, as well as for individual foresters.

Sixth, certification needs more scientific support and inputs. The need for objective indicators for the broad and ever evolving concept of sustainable forest management poses a great challenge for forestry research. Scientific support, in turn, is critical for the sustainability of certification itself. Unless certification claims can be supported by scientific evidence, they will ultimately lack credibility.

Finally, certification is no panacea for the problems of forestry. Forestry issues have intricate ramifications for many areas, ranging from pure biological components of forest ecosystem dynamics to broad intersectoral and international policy issues. However, the role of certification as a catalyst of good forest stewardship may be quite significant. As the paradigms of forestry change and different stakeholders become involved in contributing to the concepts of sustainable forestry, it is crucial to have concrete examples of what those concepts are. Certification could provide recognized cases of good forest management systems that have impacts at the local and regional levels, as catalysts of changes toward sustainable forestry.

Scenarios

Predicting the future of certification is risky. Instead we explore below three different scenarios for certification: an optimistic, a pessimistic, and a catalytic one.

Optimistic Scenario

In this scenario, forest certification becomes part of mainstream forestry for major forest-producing regions of the world within the next 10 years. This scenario implies that certification becomes supported by major forestry institutions at the national and international levels. Certification would then become a part of regular protocols of forest management, and certified operations would receive governmental and intergovernmental support. This scenario would depend on negotiations and institutional linkages within and between key international players such as Forest Stewardship Council, International Tropical Timber Organization, International Organization for Standardization, World Trade Organization, Food and Agriculture Organization Commission for Sustainable Development, and the Intergovernmental Panel on Forests. It would also depend on the internal dynamics of key producing and consuming countries.

Finally, this optimistic scenario depends on an increasing environmental awareness by consumers and a willingness to purchase "green products" to influence the market and the decision-making process at both national and international levels.

Pessimistic Scenario

Economic viability of certification schemes is vital for their credibility. There are indications that, although a significant part of the world markets has increasing environmental concerns, consumers are not willing to pay prices for certified products that are much higher than those of uncertified products. The costs of certification are now often being subsidized by various donors with the understanding that eventually certifiers and support institutions will be self-sustainable. If a time comes when certification is proven not to be economically viable either through higher prices or better market share for certified products, then it may collapse and become discredited.

The essence of certification is credibility. Credibility is more likely to be secured by independent third-party evaluations controlled by institutions with high credibility among the public, rather than by governmental institutions. Governmental institutions often suffer from credibility and institutional capacity problems at present. In fact, if governmental organizations had the necessary credibility, certification might have never appeared as an independent movement. Therefore, a significant threat to certification is excess government involvement through mandatory regulations. There is a very thin line between desirable governmental involvement and support, and inappropriate governmental control of certification.

Catalytic Scenario

Certification assumes an important role as a catalyst of change in the development of sustainable forestry. The development of criteria and indicators of sustainability at regional or national levels, as well as the development of guidelines for certification at the forest management unit level, is developing, and will continue to have profound influences. The participatory processes of consultation with multiple stakeholder groups from different geographical regions has also provided a new dimension in the global perspective of sustainable forestry. This contribution alone will probably secure an important place for certification in the history of world forestry.

In addition to its conceptual contribution to sustainable forestry, certification will undoubtedly contribute to translating theoretical ideas into practical experiences providing opportunities for dialogue, training, and research in different forest ecosystems.

Even if certification ends up merely as a transient event in forestry history, it will have made a significant contribution to good forest stewardship. Sustainable forestry, like sustainable development, is something that we cannot afford to neglect. The debate over certification has certainly contributed to our understanding of sustainable forestry.

The Editors

About the Contributors

The Editors

VIRGÍLIO M. VIANA is professor of tropical forestry at the University of São Paulo, Brazil, where he teaches graduate and undergraduate classes and conducts independent research. He holds a doctorate in biology from Harvard University, and has over ten years of experience working in tropical forest ecology and management research. He has published numerous articles in various professional journals on these subjects. He was also a senior associate at the Center for International Forestry in Bogor, Indonesia. Recently he spent a sabbatical researching and teaching forest ecology at the University of Florida.

JAMISON ERVIN studied international administration at the School for International Training in Brattleboro, Vermont. She is currently working on a doctorate in natural resources at the University of Vermont. She spent five years in Nepal studying anthropology and working for the Britain Nepal Medical Trust in community health and literacy. A member of the board of directors of the Good Wood Alliance (GWA), she has also worked for the Forest Stewardship Council since 1992, and is currently the FSC U.S. contact person.

RICHARD Z. DONOVAN is director of the Smart Wood Program at the Rainforest Alliance. He is also visiting lecturer at the University of Vermont Environmental Program. As a World Wildlife Fund senior fellow, he spent three years in Costa Rica directing the Osa Peninsula Forest Conservation and management Project. He also spent seven years as a natural resources management specialist with Associates in Rural Development, Inc., in Burlington, Vermont, and three years as a Peace Corps volunteer in Paraguay.

CHRIS ELLIOTT studied plant sciences and forestry at the University of London and Yale University. He is currently working on a doctorate in forest policy at the Swiss Federal Institute of Technology. After having worked in the field of organic agriculture and after a brief period as a research assistant at the World Bank, he has worked for WWF International in Switzerland for the past nine years. He is a former chairman of the Forest Stewardship Council Board, and serves on the Scientific Advisory Board of the European Forest Institute.

HENRY L. GHOLZ is professor of forest ecology in the School of Forest Resources and Conservation, where he teaches undergraduate and graduate courses and conducts research into the structure and dynamics of forest ecosystems in the southeastern United States, Brazil, Costa Rica, and Bolivia. He is author of over fifty research articles and has edited books in the areas of agroforestry, the carbon dynamics of pine forests, and remote sensing applications to forest productivity modeling. From 1994 to 1995 he was an international forestry advisor to the U.S. Agency for International Development as a diplomacy fellow with the American Association for the Advancement of Science.

The Chapter Authors

BRUCE CABARLE is a senior associate of the World Resources Institute and director of Forestry and Land Use in WRI's Center for International Development & Environment. His areas of expertise include forestry certification programs, forest policy and land-use planning, community forestry, forest concession management, and the analysis of forestry and agricultural projects for carbon sequestration. He has served since 1995 as chairman of the board for the Forest Stewardship Council, and is also a member of the Forestry Advisors Group—an ad hoc body of international development aid agencies concerned with the forest sector in developing countries.

MICHAEL GROVES is currently an advisor to SGS Indonesia on forest management certification and environmental management systems. He has six years experience in environmental and forest management consultancy. He is also a lead assessor for Qualifor, SGS Forestry's quality forest management certification program.

KATE HEATON is the associate director of the Rainforest Alliance's Smart Wood Program. Prior to Smart Wood, she worked for three years as a science policy advisor on international forestry and climate change for the U.S. Environmental Protection Agency (USEPA). She received a masters degree in tropical ecology from Yale School of Forestry in 1989 and studied tropical biology with the Organization for Tropical Studies (OTS) in Costa Rica.

FRANK MILLER has almost fifteen years experience of forest research and consultancy, specializing in inventory, information technology, management of forestry projects, and forest management certification and assessments. He has worked as a forest inventory specialist with British Aid and as a microcomputer advisor throughout the world. Over the last three years he has spearheaded the development of SGS Forestry's forest management certification program Qualifor and holds the position of program director.

ALAN R. PIERCE has worked as a consultant to the Forest Stewardship Council since 1992. He is a 1996 masters candidate in the forestry program at the University of Vermont's School of Natural Resources. He currently specializes in nontimber forest products research.

FRANCIS E. PUTZ is a senior associate at the Center for International Forestry Research (CIFOR) and a professor of botany and forestry at the University of Florida. His research focuses on the ecological basis of forest management for timber and nontimber forest products. In particular, he works to provide a sound scientific foundation for market-based incentives for good forest stewardship.

MARKKU SIMULA is a forest economist who has worked for FAO, academic and research institutions, and the forest industry. He has carried out consultancy assignments for international organizations, financing institutions, and bilateral agencies. Since 1980 he has been President of Indufor, a consulting company specializing in forestry and environmental management. He is also an adjunct professor of forest economics at the University of Helsinki.

TIMOTHY SYNNOTT is a forester whose experience is mainly in tropical forest management. He worked in Uganda from 1965 to 1973, and since then he has worked in over 30 tropical and subtropical countries. In 1994 he was appointed executive director of the Forest Stewardship Council.

The Stakeholder Perspective Contributors

ANTONY CARMEL is executive director of Soltrust, a nongovernmental organization dedicated to promoting sustainable forest management and community development in the Solomon Islands.

JULIO CÉSAR CENTENO is an independent consultant and a university professor of wood technology.

RACHEL CROSSLEY formerly worked at the World Bank on forest-related issues. She is currently working with Environmental Advantage in New York City.

MARK EISEN is director of environmental marketing for The Home Depot, a large chain of home improvement stores in the United States.

ANNA FANZERES formerly worked with Greenpeace Brazil. She is currently working toward a doctorate at Yale University.

DEBBIE HAMMEL is director of the Forest Conservation Program at Scientific Certification Systems in Oakland, California.

HOWARD HEINER worked for the International Union of Societies of Foresters.

JIM HODKINSON works for B & Q, a large chain of hardware and home repair stores in the United Kingdom.

BILL HOWE is a forester at Collins Pine, one of the first certified forest operations in the United States.

PAULA ROGERS HUFF is director of the Sustained Development Institute, College of Menominee Nation.

MICHAEL JENKINS is associate director of the World Environment and Resources Program at the John D. and Catherine T. MacArthur Foundation in Chicago.

RUSSEL JOHNSON works in environmental affairs at IKEA Furniture, the world's largest furniture manufacturer.

SCOTT LANDIS is the founder of WARP (Good Wood Alliance), a nongovernmental organization promoting conservation of forest resources through responsible wood use.

GERALD LAPOINTE works for the Canadian Pulp and Paper Association in Montreal, Quebec.

WILLIAM E. MANKIN is director of the Global Forest Policy Project, a joint project of the Sierra Club, the National Wildlife Federation, and Friends of the Earth U.S.

JOHN P. MCMAHON works with Weyerhauser International in Tacoma, Washington.

SIAN TUAN MOK is an independent sustainable forestry consultant based in Malaysia.

ERIC PALOLA is a resource economist at the National Wildlife Federation's Northeast Natural Resource Center in Montpelier, Vermont.

MARSHALL PECORE is the chief forest manager at Menominee Tribal Enterprises, a corporation which manages 220,000 acres of forest in Wisconsin.

ROBERT WAFFLE works with International Wood Products Association in Alexandria, Virginia.

Index